Refrigerators, Heat Pumps and Reverse Cycle Engines

SCIENCES

Energy, Field Directors – Alain Dollet and Pascal Brault

Physics of Energy and Energy Efficiency,
Subject Head – Michel Feidt

Refrigerators, Heat Pumps and Reverse Cycle Engines

Principles, State of the Art and Trends

Coordinated by
Jocelyn Bonjour

WILEY

First published 2023 in Great Britain and the United States by ISTE Ltd and John Wiley & Sons, Inc.

ISTE Ltd
27-37 St George's Road
London SW19 4EU
UK

www.iste.co.uk

John Wiley & Sons, Inc.
111 River Street
Hoboken, NJ 07030
USA

www.wiley.com

Library of Congress Control Number: 2022947623

British Library Cataloguing-in-Publication Data
A CIP record for this book is available from the British Library
ISBN 978-1-78945-123-8

ERC code:
PE8 Products and Processes Engineering
PE8_6 Energy processes engineering

Contents

Foreword

Michel FEIDT
Université de Lorraine, LEMTA, CNRS, Nancy, France

The book you have in your hands is one of the books from the "Physics of Energy and Energy Efficiency" subject in the Engineering and Systems department.

The subject of "Physics of Energy and Energy Efficiency", albeit recent, is not new. It is particularly underpinned by a thermodynamic approach, whatever the scale.

The selected aspect will be phenomenological and characterized explicitly in order to emphasize the key concept of "efficiency", essential for any system or process.

The characterization chosen for the development of this subject has been arranged into four successive books, each strongly correlated with each other, and also with other series within the department:

– *Fundamental Physics of Energy*;

– *Thermodynamics of Heat Engines*;

– *Heat Engines with Inverse Cycles*;

– *Efficiency in Practice*.

I would like to thank ISTE, the various coordinators, and the authors for their contributions and effective actions, despite the very particular conditions of the moment. We are awaiting and would like to encourage comments, suggestions and questions from readers.

Preface

Jocelyn BONJOUR
CETHIL, INSA Lyon, Villeurbanne, France

For thousands of years, the quest for energy has been focused on a single purpose: the production of heat (thermal energy), mainly for domestic heating and cooking. Combustion, which practically used to be the only energy conversion technology, was gradually mastered and improved.

Combustion was still at the heart of the Industrial Revolution, which was triggered when heat could be used to produce mechanical work. The title of Sadi Carnot's book is unambiguous: the founder of thermodynamics shared in 1824 his *Reflections on the Motive Power of Fire and on Machines Fitted to Develop that Power*. It was during this period that many thermo-mechanical energy conversion systems were developed. These systems had in common that they were based on cyclic transformations of fluids, then called "engine cycles" or "direct cycles". Thermodynamicists then considered reversing the engine cycle: reverse cycle engines were born, used as refrigerators, heat pumps or even double-function heat pumps.

Climate change and the energy crisis have undoubtedly called for real changes in energy paradigms. Engine cycles will still be widely used to produce mechanical and electrical work, provided that the thermal energy sources are powered less and less by the combustion of fossil fuels. Similarly, the share of thermal energy production derived directly from combustion is bound to decrease over time. Finally, due to global warming, the need for cooling is bound to increase, whether for refrigeration for food safety and sanitary purposes, or for the cooling of premises. It is therefore clear that, more than ever, reverse cycle thermal engines will play an important role in the energy landscape of the coming decades.

The objective of this book is to offer readers (graduate students, PhD students, engineers, researchers) a state of the art on reverse cycle engines, in order to prepare them for their future positions, as well as to outline the research trends on still emerging technologies.

Thus, Chapter 1 presents a scientific and technical state of the art concerning heating and cooling by the most common reverse cycle engines: vapor compression, ejection, absorption or adsorption engines.

The energy crisis invites us to improve the energy efficiency of systems, whose performances must be evaluated with rigor and precision. This is the purpose of Chapter 2, which develops entropy and exergy analysis methods applied to reverse cycle engines and to the scale of their components. Chapter 3 completes the approach by proposing optimization methods of systems, by way of finite time/finite speed thermodynamics.

The development of mechanical compression engines during the 20th century has made this technology very mature, so that the margins for progress are less than for thermal compression engines (absorption, adsorption), which are therefore the subject of various research projects. There are still some scientific and technical obstacles which are discussed in Chapter 4, as well as the avenues envisaged to overcome them.

Magnetic refrigeration is an emerging technology. It is based on a reverse cycle like the systems mentioned above, but it is a magnetic material (and not a fluid) that undergoes a set of cyclic transformations. Chapter 5 presents the principle of this technology and different current or future applications.

Finally, Chapter 6 presents the thermoelectric effect as an alternative to reverse cycle engines. A good understanding of this physical phenomenon allows us to analyze the performance of thermoelectric refrigeration systems and to identify some applications for which they could be particularly relevant.

We hope that this book will enlighten the reader on the operation and future evolution of all the reverse cycle engines used for heating and cooling, as well as on their essential role in the decades to come.

Lyon
March 2023

1

Heating and Cooling by Reverse Cycle Engines: State of the Art

Philippe HABERSCHILL and Rémi REVELLIN
CETHIL, INSA Lyon, Villeurbanne, France

Heat pumps are, from a thermodynamic point of view, no different from refrigerators: in both cases, they are a thermal generator which, thanks to energy consumption, passes heat from a cold source to a hot source (heat sink). On the one hand, the purpose of heat pumps is to provide heat to the hot source (air of a building, domestic hot water, swimming pool, etc.). On the other hand, refrigerating machines make it possible to obtain and maintain a system at a temperature lower than the ambient temperature. To do this, it is necessary to remove heat from this system or even "produce cold". There are two main types of reverse cycle thermal generators: vapor compression systems (two-heat-source systems) and systems driven by thermal energy (three-heat-source systems). This chapter presents different configurations of vapor compression systems to describe two common systems driven by thermal energy: absorption systems and ejection systems.

For a color version of all the figures in this chapter, see www.iste.co.uk/bonjour/refrigerators.zip.

Refrigerators, Heat Pumps and Reverse Cycle Engines,
coordinated by Jocelyn BONJOUR.

1.1. Vapor compression refrigerators and heat pumps

Air refrigeration systems were the first compression refrigerators used. They are increasingly being abandoned (except in particular in the field of very low temperatures: cryogenics) in favor of vapor compression systems, and therefore of condensable fluids under the conditions of use. Such machines, thanks to the use of the refrigerant latent heat of change of state, make it possible to obtain refrigerating effects per unit mass of fluid that are clearly superior to those of gas systems. The systems are thus smaller in size.

1.1.1. *Operation principle of closed-circuit refrigeration installation: definitions*

As in heat engines, the system considered is a fluid in cyclic evolution. This fluid, which is intended to exchange heat with the sources, is called the refrigerant.

If the fluid absorbs the amount of heat Q_f from the cold source (CS), it therefore releases Q_c to the hot source (HS) (Figure 1.1):

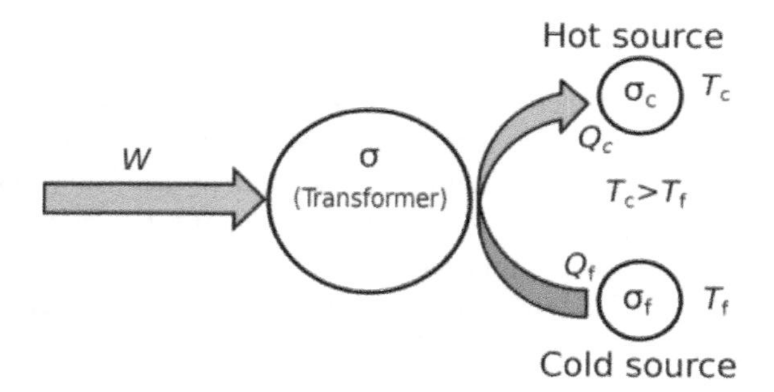

Figure 1.1. *Thermal generator*

$$|Q_c| = Q_f + W \quad [1.1]$$

where W is the mechanical (or other) energy received.

NOTE.– The purpose of a refrigerator is to extract Q_f from the cold source, whereas the purpose of a heat pump is to deliver $|Q_c|$ to the hot source. Essentially, these two systems are no different.

The amount of heat Q_f taken at the cold source is called the refrigeration effect or cooling capacity.

The ratio $\varepsilon = \frac{Q_f}{W}$ is called the energy efficiency ratio or coefficient of performance.

Let us look for the expression of ε in two different cases: reversible operation and irreversible operation.

1.1.1.1. *Reversible operation*

Like a heat engine, a thermodynamic generator can operate reversibly (internal and external reversibilities) between two thermal sources only if the evolution cycle of the refrigerant is a Carnot cycle. T_c is the maximum temperature and T_f is the minimum temperature (Figure 1.2).

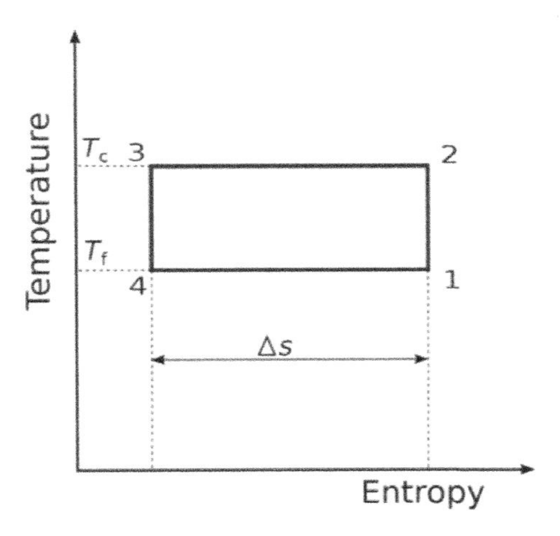

Figure 1.2. *Carnot engine*

Heat and work per unit mass of fluid will be denoted by q and w, respectively, expressed in J/kg

$$q_f = T_f \Delta s,\ q_c = T_c \Delta s \tag{1.2}$$

and

$$w = |q_c| - q_f = (T_c - T_f)\Delta s \tag{1.3}$$

Therefore:

$$\varepsilon_{Carnot} = \frac{T_C}{T_C - T_f} \quad [1.4]$$

with s being the specific entropy of the fluid expressed in J/kg.K. Moreover, ε_{Carnot} is the Carnot energy efficiency ratio. Thus, ε_{Carnot} can be greater than 1 depending on the value of T_f with respect to $T_c - T_f$:

$\varepsilon_{Carnot} > 1\ if\ T_f > T_c - T_f \leftrightarrow T_f > \frac{T_C}{2}$ is the most frequent case

$\varepsilon_{Carnot} < 1\ if\ T_f < T_c - T_f \leftrightarrow T_f < \frac{T_C}{2}$; this is the case for the liquefaction of certain gases.

1.1.1.2. *Irreversible operation*

For any vapor compression refrigerating machine, the quantities of heat are given by:

$$q_f = T_f|\Delta s_f| \text{ and } q_c = T_c|\Delta s_c| \text{with } |\Delta s_c| = |\Delta s_f| + s'$$

where s' represents the entropy generation.

The mechanical energy exchanged is therefore: $w = (T_c - T_f)|\Delta s_f| + T_c s'$

So

$$\varepsilon = \frac{T_C}{(T_C - T_f) + T_C \frac{s'}{\Delta s_f}} \quad [1.5]$$

For refrigeration machines, as for engines, the Carnot cycle leads to the highest energy efficiency ratio. The difference between any machine and a Carnot machine is measured by the cooling efficiency η_f, which by definition is:

$$\eta_f = \frac{\varepsilon}{\varepsilon_{Carnot}} \quad [1.6]$$

η_f also corresponds to the exergy efficiency η_{ex} of the refrigeration cycle if the reference temperature T_{ref} is considered equal to that of the hot source T_c. The exergy efficiency is therefore a better indicator of the quality of the thermodynamic cycle than the energy efficiency ratio, in that it is immediately interpreted as a ratio of the actual performance to the ideal performance.

A heat pump is, from a thermodynamic point of view, no different from a refrigerator: in both cases, it is a thermal generator which, thanks to energy consumption, transports heat from a cold source to a hot source. Thus, the previous systems, qualified as "refrigerators", can all be used as a heat pump. The difference between these two types of machines lies in how they are used. What is interesting about a heat pump is the quantity of heat q_c which will be supplied to the hot source. This difference of interest gives the definition of the coefficient of performance (COP):

$$COP = \frac{|q_c|}{w} \quad [1.7]$$

This relation is, according to the first law ($|q_c| = q_f + w$), always greater than 1, meaning that these systems are of great theoretical and practical interest. Indeed, unlike other heating processes, this one makes it possible to obtain thermal energy greater than the energy expended to obtain it. The difference of course comes from the energy "pumped" into the cold source.

With the theoretical and technological development of heat pumps being modeled on that of refrigerating systems, there is no need to repeat it here. Emphasis will simply be placed on the difference between the definitions of the energy efficiency ratio ε, on the one hand, and of the COP, on the other hand, which of course leads to differences in expressions. For example, the COP of a heat pump operating according to the Carnot cycle is:

$$COP_{Carnot} = \frac{T_c}{T_c - T_f} \quad [1.8]$$

The relative coefficient of performance corresponds to the exergy efficiency η_{ex} (if the reference temperature is equal to the temperature of the cold source) of a heat pump and is given by:

$$COP_{relative} = \frac{COP}{COP_{Carnot}} \quad [1.9]$$

1.1.2. *Actual cycle with superheating and subcooling*

Figure 1.3 shows a diagram of a refrigerator (or heat pump), as well as the evolution cycle of the associated refrigerant. At state 1, the vapor, at low pressure, is either saturated (vapor quality of 1) or superheated (as in the example). The vapor is then compressed in a compressor where its pressure and temperature are increased (point 2). The superheated vapor is then condensed in a condenser at the outlet of which the fluid is in a liquid state, either saturated (vapor quality equal to 0) or subcooled (point 3). The liquid is then expanded in an expansion valve and partial vaporization is observed (vapor quality of around 0.2–0.3 at point 4). This two-phase fluid is then vaporized in an evaporator to reach state 1.

It should be noted that for a domestic refrigerator, the cold source corresponds to the refrigerated enclosure, while the hot source is represented by the air in the kitchen. Conversely, for a residential heat pump, the cold source and the hot source correspond respectively to the outside air and to the fluid to be heated (air, domestic hot water (DHW), etc.) inside the building.

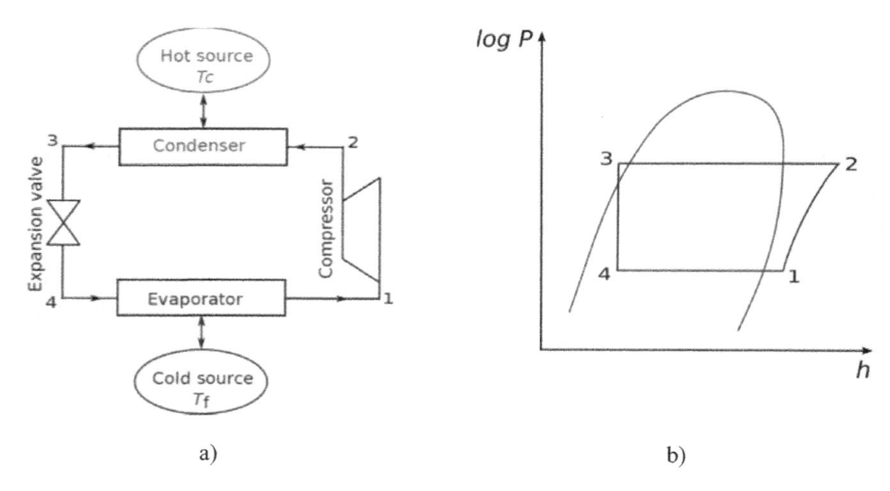

Figure 1.3. *Classic cycle of a vapor compression refrigerator or heat pump. a) Schematic of the installation. b) Cycle in an enthalpy diagram*

1.1.3. *Special cycles*

1.1.3.1. *Transcritical cycles*

A pressure higher than the critical pressure of the fluid in the high-temperature exchanger can be a particularity encountered in certain vapor compression cycles. Such cycles are called transcritical. Refrigeration machines using CO_2 as a refrigerant are often transcritical when the ambient temperature is higher than the critical temperature of the fluid. There is then no more condensation in the "hot" exchanger, but a cooling of the gas. An example of a transcritical cycle is shown in Figure 1.4, which represents a refrigeration cycle. In this figure, there is a strong change in temperature in the high-pressure exchanger: from approximately 120°C to 30°C. There are strong thermal differences between the source, whose temperature varies little, and the refrigerant in the exchanger, which leads to strong transfer irreversibility and contributes to a deterioration in the efficiency in this type of system.

Nevertheless, this strong temperature gradient on the refrigerant can become an advantage in the case of strong variations in the temperature of the hot source, such as in heat pumps intended to produce DHW. In this case, the "hot source", which is the water to be heated, has a temperature which must change from the network temperature (generally below 20°C) to a temperature above 60°C. The temperature glide of the refrigerant is then partly compensated by the source temperature glide, which reduces irreversibility.

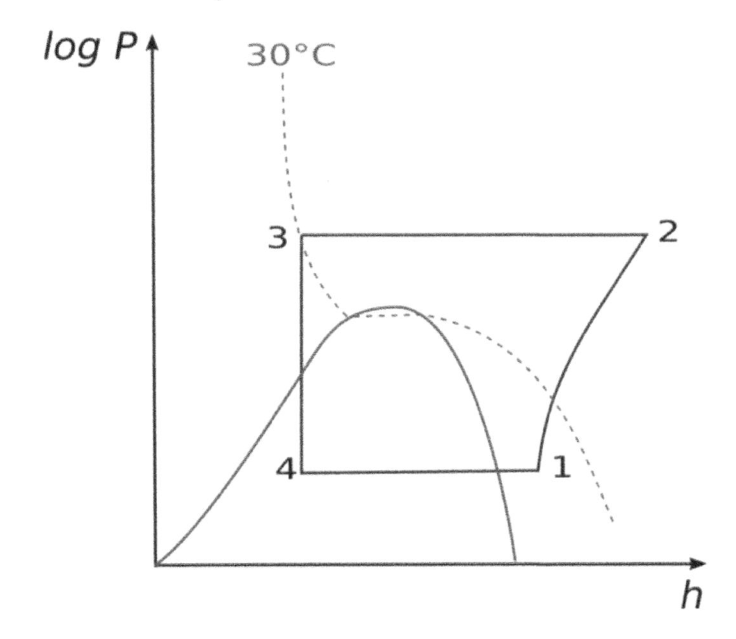

Figure 1.4. *Example of transcritical cycle (case of CO_2)*

Note that the HP (high pressure) is no longer conditioned by the condensing temperature, but is left to the discretion of the designer. However, at a constant outlet temperature, there is an optimum pressure:

– below the critical pressure, the refrigeration production is zero because the entire cycle is in the vapor phase;

– at the critical pressure, the cooling capacity is low due to a minimum enthalpy variation in the evaporator;

– for pressures above the critical pressure, the cooling production increases;

– if the pressure is too high, the gain in cooling production is less than the additional cost of compression and the coefficient of performance deteriorates.

The optimum pressure depends on the operating conditions, but is approximately 90 bar for an air-cooled gas cooler with an outlet temperature of 35°C and an evaporation temperature of 0°C.

1.1.3.2. *Multistage cycles*

In certain thermal situations, or for certain refrigerants, the overall compression ratio requires compression to be carried out in two stages, or even more. The fluid passes from one stage to another to cool it in an intermediate exchanger (direct contact or mixer).

Several machine construction schemes are then possible. In all these examples, the compressions will be considered adiabatic. As a result, the compression works will be expressed simply from a difference in enthalpy.

1.1.3.2.1. Heat exchanges with an external source

This system is represented in Figure 1.5(a) with its associated cycle (Figure 1.5(b)). The fluid leaving the LP stage passes through an exchanger connected to the external source at temperature T_M before entering the HP stage. The sum of the specific work of both stages is clearly lower than the specific work for a single stage. The energy efficiency ratio is expressed by:

$$\varepsilon = \frac{h_1 - h_4}{h_a - h_1 + h_2 - h_b} \quad [1.10]$$

The COP of the installation, meanwhile, is written as:

$$COP = \frac{h_3 - h_2}{h_a - h_1 + h_2 - h_b} \quad [1.11]$$

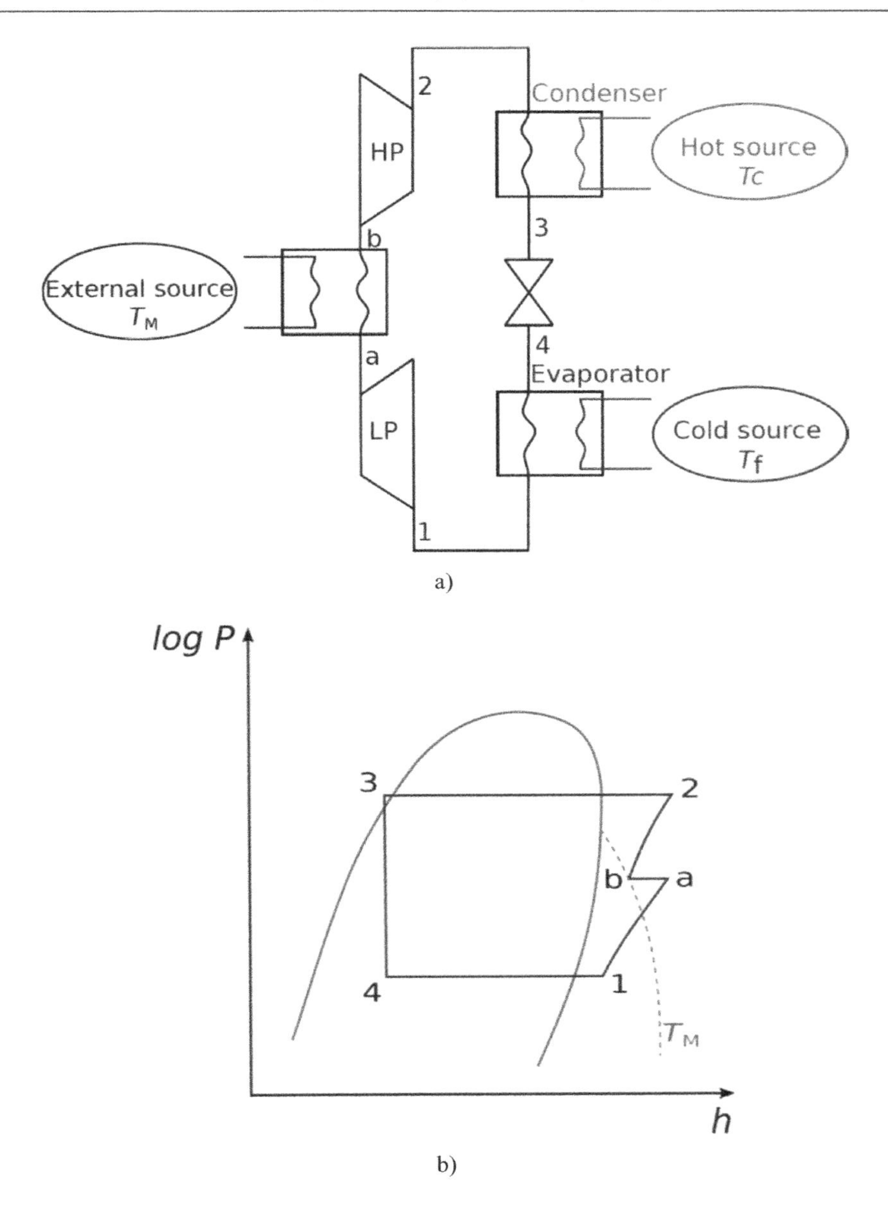

Figure 1.5. *Two-stage compression: heat exchange with an external source. a) Diagram of the installation. b) Cycle in an enthalpy diagram*

1.1.3.2.2. Heat exchanges with the evaporating fluid

In Figure 1.6(a), the fluid leaving the LP stage is cooled by the cold fluid coming from the expansion of a fraction (f) of the fluid leaving the condenser. The energy efficiency ratio is expressed by:

$$\varepsilon = \frac{(1-f)(h_1-h_4)}{h_a-h_1+h_2-h_b} \quad [1.12]$$

with the fraction f of fluid removed, which can be determined thanks to an enthalpy balance on the exchanger:

$$f = \frac{h_a-h_b}{h_1-h_4} \quad [1.13]$$

Note that in this case $h_1 \approx h_{1'} \approx h_{1''}$ and that $h_4 \approx h_{4'}$.

The COP of the installation, meanwhile, is written as:

$$COP = \frac{(h_3-h_2)}{h_a-h_1+h_2-h_b} \quad [1.14]$$

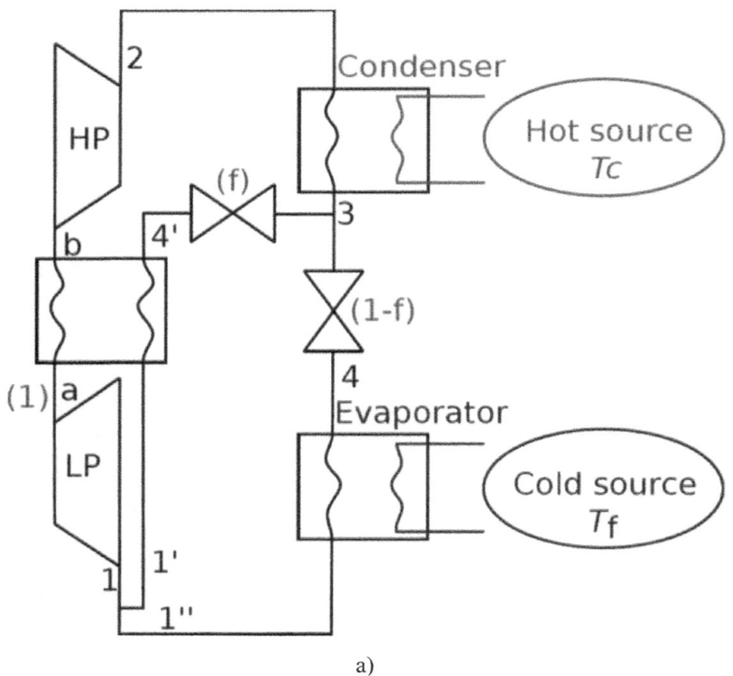

a)

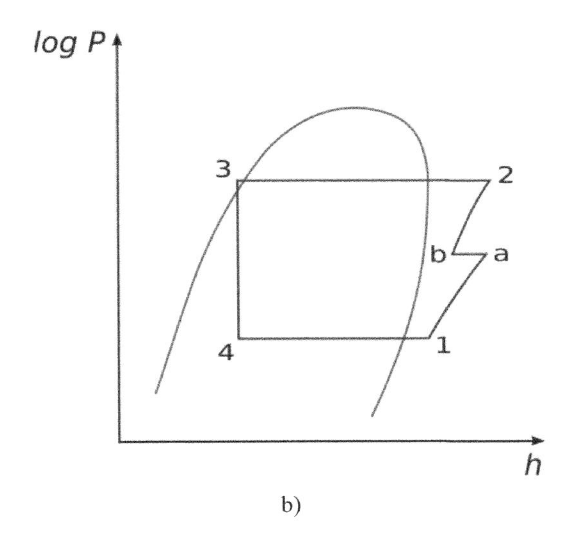

b)

Figure 1.6. *Two-stage compression: heat exchange with the evaporating fluid. a) Diagram of the installation. b) Evolution cycle of the fluid in an enthalpy diagram. Note that in this case* $h_1 \approx h_{1'} \approx h_{1''}$ *and* $h_4 \approx h_{4'}$

1.1.3.2.3. Injection into an intermediate reservoir

In Figure 1.7(a), the fluid leaving the LP stage is cooled by the evaporation (endothermic reaction) of the fraction of the fluid (f) leaving 4'. At point b, the fluid is in the form of saturated vapor because it is removed from the upper part of the reservoir containing the two-phase refrigerant in the liquid-vapor state. The evolution cycle in the enthalpy diagram is given in Figure 1.7(b). The energy efficiency ratio is expressed by:

$$\varepsilon = \frac{(1-f)(h_1-h_4)}{(1-f)(h_a-h_1)+h_2-h_b} \quad [1.15]$$

The fraction f of fluid removed can be determined thanks to an enthalpy balance on the intermediate reservoir:

$$f = \frac{h_a-h_b}{h_a-h_3} \quad [1.16]$$

The COP of the installation, meanwhile, is written as:

$$COP = \frac{(h_3-h_2)}{(1-f)(h_a-h_1)+h_2-h_b} \quad [1.17]$$

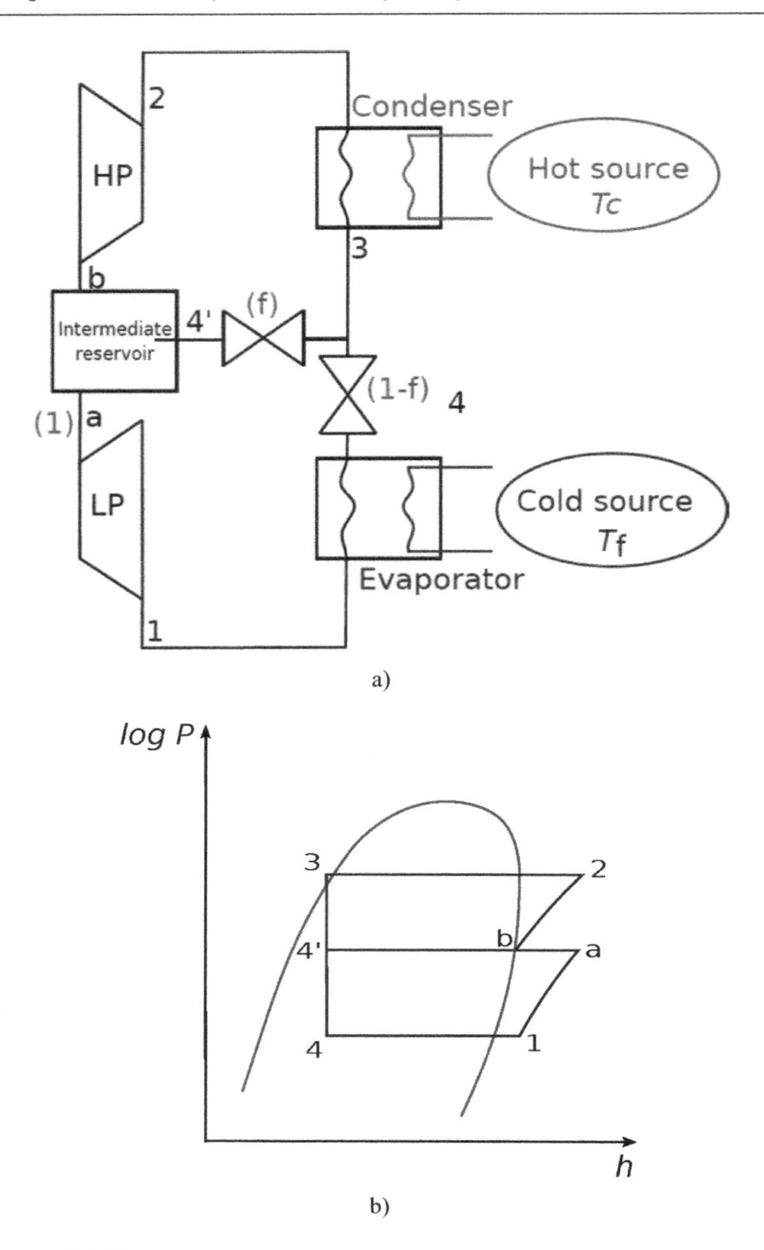

Figure 1.7. *Two-stage compression: injection into an intermediate reservoir. a) Diagram of the installation. b) Evolution cycle of the fluid in enthalpy diagram*

1.1.3.2.4. Injection into an intermediate reservoir with staged expansion

An intermediate *mixer*-type heat exchanger (Figure 1.8(a)) at intermediate pressure P_i receives the superheated vapor, which leaves the low-pressure (LP) stage of the compressor in state a (Figures 1.8(b)), and the fluid from the high-pressure (HP) expansion device in a two-phase state c. The dry saturated vapor (state b) leaves the mixer to enter the HP stage of the compressor. The saturated liquid, which leaves the bottom of the mixer in state d, feeds the LP expansion valve which supplies the evaporator with a fluid in state 4.

For such a refrigerating system, the energy efficiency ratio is given by:

$$\varepsilon = \frac{(1-f)(h_1-h_4)}{(1-f)(h_a-h_1)+h_2-h_b} \qquad [1.18]$$

with the fraction f of fluid removed, which can be determined thanks to an enthalpy balance on the mixer:

$$f = \frac{h_d-h_a-h_c+h_b}{h_d-h_a} \qquad [1.19]$$

The COP of the installation, meanwhile, is written as:

$$COP = \frac{(h_3-h_2)}{(1-f)(h_a-h_1)+h_2-h_b} \qquad [1.20]$$

A possible improvement of this system in terms of energy consists of precooling the fluid, leaving the LP stage (state a) by the hot fluid.

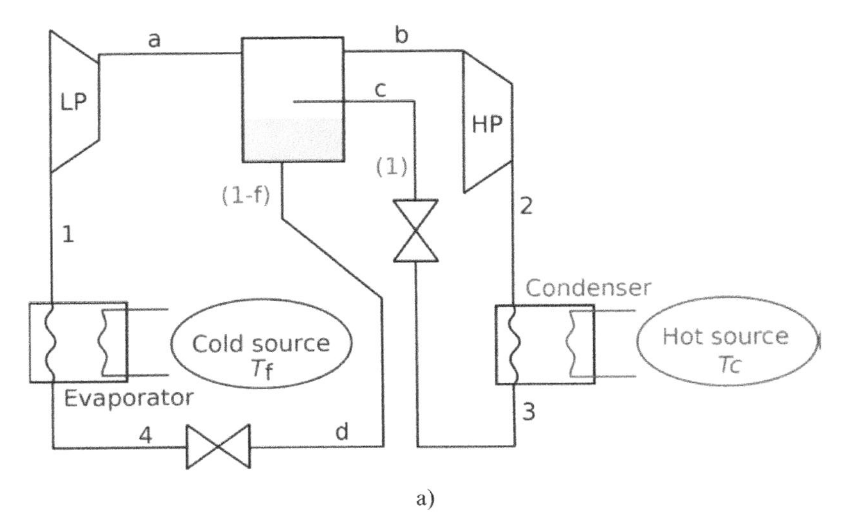

a)

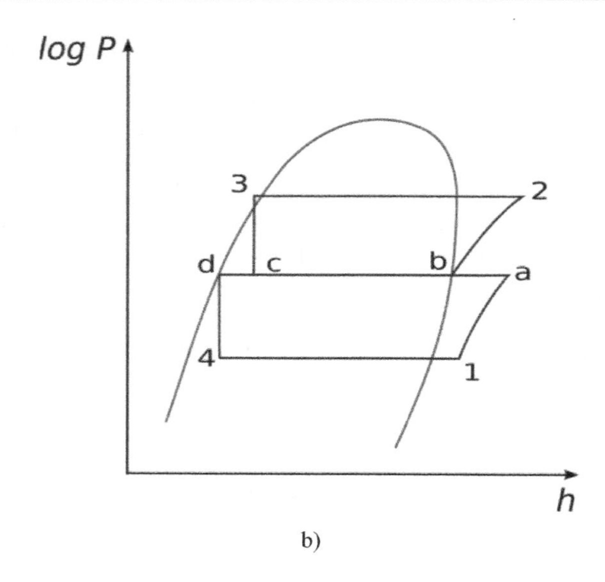

b)

Figure 1.8. *Diagram of a two-stage compression and expansion refrigerating machine and cycles of evolution of the refrigerant. a) Diagram of the installation. b) Evolution cycle of the fluid in an enthalpy diagram*

1.1.3.2.5. Cascade cycles

The use of a pure phase change refrigerant remains limited to the temperature interval between the critical point temperature, at which the latent heat of transformation is canceled out, and the triple point temperature, below which any simple mechanical cycling disappears. Moreover, this same temperature difference would cause excessive technical constraints to appear, mainly linked to the volume of the fluid at low pressure, to the difference between high and low pressures, hence correlatively amplified irreversibility.

A judiciously chosen cascade of fluids is then naturally used to ensure lower and lower temperature levels under reasonable pressures. Figure 1.9 shows the block diagram of such an installation in the case of a two-stage cascade. The extension to a higher number of stages is done by iteration.

Currently, two-stage cascade machines are used with CO_2 in commercial refrigeration, especially in supermarkets as units that can be combined in two different ways:

– CO_2 for both stages (low and high temperatures). In this case, the high-temperature stage is very often transcritical;

– CO_2 for the low-temperature stage combined with propane or ammonia or even a low GWP HFO for the high-temperature stage.

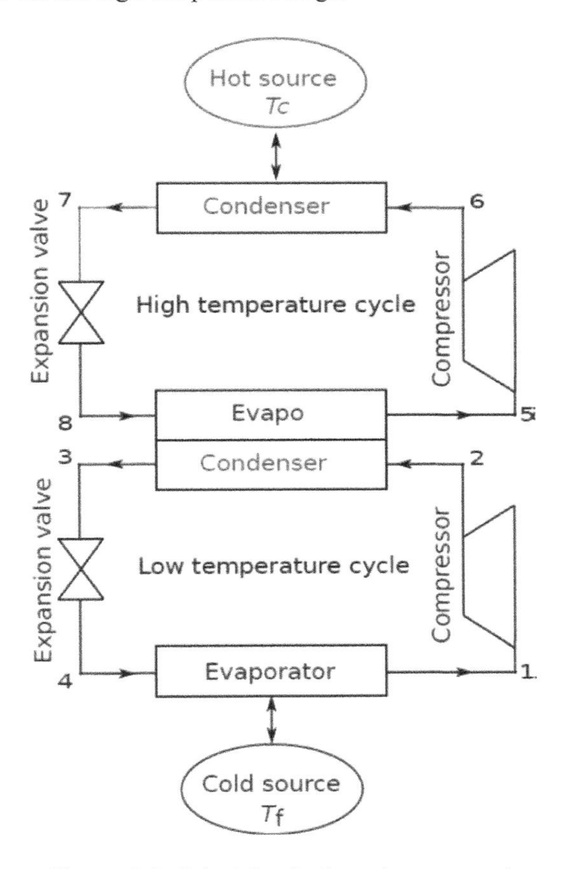

Figure 1.9. *Principle of a two-stage cascade*

1.1.3.2.6. Ejection cycles

An ejector can be used in vapor compression systems, as well as in systems driven by thermal energy (section 1.2.3). The former is a component that allows part of the expansion energy to be recovered in order to suck up the fluid coming from the evaporator. This has the effect of reducing the compression ratio at the terminals of the compressor and reducing the energy required for compression.

The ejector is depicted in Figure 1.10. It uses the Venturi effect of a convergent–divergent nozzle (if sonic at the throat), or simply a convergent nozzle (if subsonic at the convergent outlet): the high-pressure working fluid is accelerated in the nozzle, which induces a drop in pressure allowing, at the nozzle outlet, suction of the secondary fluid. Mixing occurs in the mixing zone before the mixture is recompressed in the nozzle outlet to the intermediate pressure.

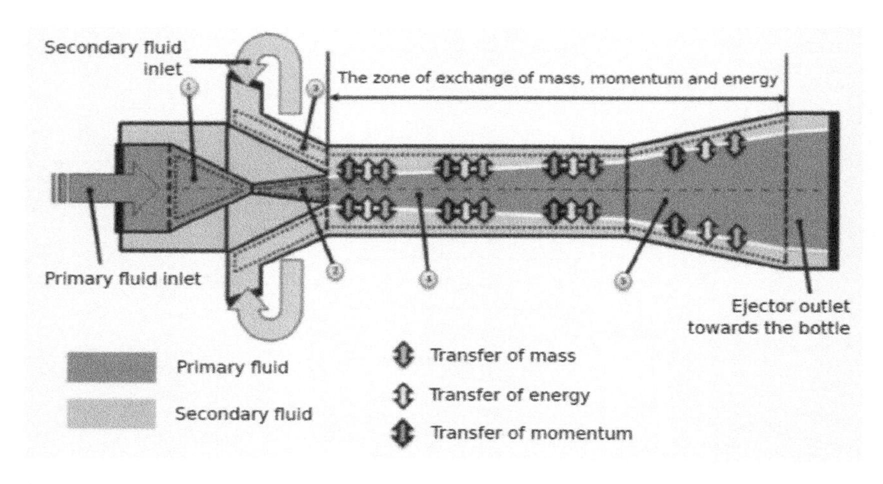

Figure 1.10. *Diagram of an ejector. 1, convergent; 2, divergent; 3, secondary fluid inlet; 4, mixing zone; and 5, nozzle for recompression (Bouziane 2014)*

There are various solutions for using an ejector to improve the operation of refrigerators, air conditioners or heat pumps.

Figure 1.11(a) shows an example of a refrigeration system operating on CO_2 according to a transcritical cycle with an intermediate exchanger. The evolution cycle of the corresponding fluid is shown in Figure 1.11(b).

The increase in the energy efficiency ratio can be as much as 20%. These ejection systems are increasingly widespread, especially for CO_2 and for various ranges of flow rates.

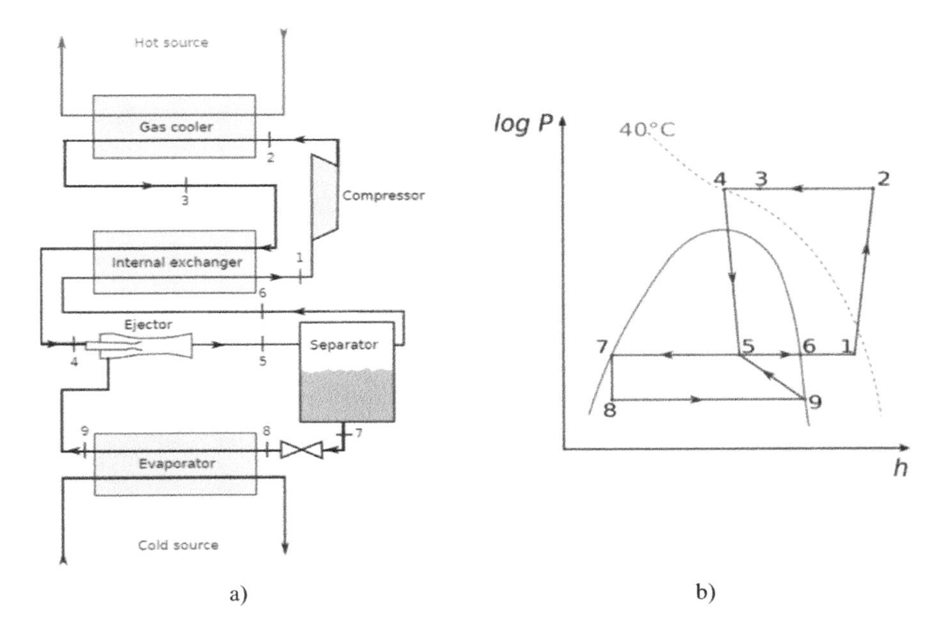

Figure 1.11. *Example of an ejection vapor compression refrigerating machine. a) Diagram of the installation. b) Evolution cycle of the fluid in an enthalpy diagram with the 40°C isotherm as an example*

1.1.3.3. *Application example*

We propose comparing three cases for the intermediate cooling of a refrigeration installation using propane as a refrigerant. The evaporation temperature is -40°C. There is a slight superheating, down to -37°C, before the compression is carried out in two stages. The outlet pressure of the LP body is 3.5 bar, and the isentropic efficiency (η_S) of each compression is 0.7. Condensation of propane ends, without subcooling, at 35°C. For these three cases, the fraction (f) is calculated and the energy efficiency ratio is compared.

Case no. 1. Surface exchanger between a fraction (f) of the fluid removed from the condenser outlet and the fluid leaving the LP body (Figure 1.6(a)). For the points numbered b and 1', the vapor quality is equal to 1.

Case no. 2. Injection into the intermediate reservoir of part (f) of the fluid taken from the condenser outlet (Figure 1.7(a)). For point b, the vapor quality is equal to 1.

Case no. 3. Staged reduction in pressure (Figure 1.8(a)).

	1	1'	a'	a	b	2'	2	3	4
P (bar)	1.11	1.11	3.5	3.5	3.5	12.18	12.18	12.18	1.11
T (°C)	-37.00	-40.00	3	16.17	-9.60	40.85	52.71	35.00	-40.00
x (-)		1			1			0	0.443
h (kJ/kg)	532.85	528.48	585.14	607.55	564.12	622.58	647.63	292.84	292.84
s (kJ/kg.K)	2.4619		2.4619		2.3840	2.3840			

Table 1.1. *Thermodynamic properties at different cycle points for propane*

In order to obtain Table 1.1, the enthalpies at the output of each compression must be calculated from the isentropic output (points a' or 2') and the isentropic efficiency

$$h_a = h_1 + \frac{h_{a'} - h_1}{\eta_s} = 532.85 + \frac{585.14 - 532.85}{0.7} = 607.55\ kJ/kg$$

$$h_2 = h_b + \frac{h_{2'} - h_b}{\eta_s} = 564.12 + \frac{622.58 - 564.12}{0.7} = 647.63\ kJ/kg$$

It is also possible to calculate the vapor quality at point 4 with the lever rule:

$$x_4 = \frac{h_4 - h_{1,4}}{h_{v,4} - h_{1,4}} = \frac{292.84 - 105.12}{528.48 - 105.12} = 0.443$$

Case no. 1

Fraction f is obtained using an enthalpy balance on the heat exchanger

$$h_a + f \cdot h_4 = h_b + f \cdot h_{1'}$$

$$f = \frac{h_a - h_b}{h_{1'} - h_4} = \frac{607.55 - 564.12}{528.48 - 292.84} = 0.184$$

The energy efficiency ratio is calculated by the following equation:

$$\varepsilon = \frac{(1-f) \cdot \Delta h_{41''}}{\Delta h_{1a} + \Delta h_{b2}} = \frac{(1 - 0.184) \times (533.84 - 292.84)}{607.55 - 532.85 + 647.63 - 564.12} = 1.24$$

For point 1'', an enthalpy balance must be carried out on the mixer:

$$h_{1''} = \frac{h_1 - f h_{1'}}{1-f} = \frac{532.85 - 0.184 \times 528.48}{1 - 0.184} = 533.84\ kJ/kg$$

This allows confirmation of the initial hypothesis that $h_1 \approx h_{1'} \approx h_{1''}$.

Case no. 2

Fraction f is obtained using an enthalpy balance on the intermediate reservoir (we note that $h_{4'} = h_4$)

$$\begin{aligned}(1-f) \cdot h_\mathrm{a} + f \cdot h_5 &= h_\mathrm{b} \\ f &= \frac{h_\mathrm{a} - h_\mathrm{b}}{h_\mathrm{a} - h_5} \\ &= \frac{607.55 - 564.12}{607.55 - 292.84} \\ &= 0.138\end{aligned}$$

The energy efficiency ratio is calculated by the following equation:

$$\begin{aligned}\varepsilon &= \frac{(1-f) \cdot \Delta h_{41}}{(1-f) \cdot \Delta h_{1a} + \Delta h_{b2}} \\ &= \frac{(1 - 0.138) \times (532.85 - 292.84)}{(1 - 0.138) \times (607.55 - 532.85) + 647.63 - 564.12} = 1.4\end{aligned}$$

Case no. 3

Fraction f is obtained by an enthalpy balance on the intermediate reservoir knowing that $h_c = h_3$. Moreover, due to the removal of the saturated liquid, the following is obtained, with $h_\mathrm{d} = 176.35\ \mathrm{kJ/kg}$:

$$\begin{aligned}(1-f) \cdot h_\mathrm{a} + h_\mathrm{c} &= h_\mathrm{b} + (1-f) \cdot h_\mathrm{d} \\ f &= \frac{h_\mathrm{a} - h_\mathrm{b} + h_\mathrm{c} - h_\mathrm{d}}{h_\mathrm{a} - h_\mathrm{d}} \\ &= \frac{607.55 - 564.12 + 292.84 - 176.35}{607.55 - 176.35} \\ &= 0.371\end{aligned}$$

The energy efficiency ratio is calculated by the following equation:

$$\begin{aligned}\varepsilon &= \frac{(1-f)\cdot \Delta h_{d1}}{(1-f)\cdot \Delta h_{1a} + \Delta h_{b2}}\\ &= \frac{(1-0.371)\times(532.85-176.35)}{(1-0.371)\times(607.55-532.85)+647.63-564.12}\\ &= 1.72\end{aligned}$$

Note that the greater the complexity, the more the energy efficiency ratio increases. Now, comparing it with a single-stage compression, it becomes:

$$\begin{aligned}s_{2'} &= s_1 = 2.4619 \text{ kJ/kg}\cdot\text{K}\\ h_{2'} &= 647.49 \text{ kJ/kg}\\ h_2 &= h_1 + \frac{h_{2'} - h_1}{\eta_s} = 532.85 + \frac{647.49-532.85}{0.7} = 696.62 \text{ kJ/kg}\\ w_t &= h_2 - h_1 = 696.62 - 532.85 = 163.77 \text{ kJ/kg}\\ \varepsilon &= \frac{h_1 - h_4}{w_t} = \frac{532.85-292.84}{163.77} = 1.47\end{aligned}$$

This theoretical result shows that compression in a single stage leads to a very good energy efficiency ratio. However, in reality, the flow rate of the compressor (volumetric) will drop because of the reduction in its volumetric efficiency when the compression ratio increases. Thus, the cooling capacity produced will be insufficient.

Compressing in two stages with intermediate cooling is therefore essential for technological reasons. The final choice of system configuration will depend on the size of the machine. For small powers (inexpensive installation), priority should be given to the cooling capacity and the value of the energy efficiency ratio will be of little importance. On the other hand, at high power (expensive installation), a certain quality of operation and a good energy efficiency ratio will be sought to ensure a certain return on investment.

1.1.4. *Heat output settings*

The cooling capacity values of a refrigerating system, or the heating capacity of a heat pump, used for the calculation and the choice of the necessary equipment, generally correspond to the maximum thermal load of the machine. Indeed, the installation must make it possible, in the most difficult conditions, to satisfy the needs for heating or cooling. When, under the action of various factors, these needs decrease, the refrigeration or heat production of the machine must be lowered to adapt to these new load conditions.

Various methods can be used to ensure the reduction of the capacity produced by these systems.

There are two solutions to reduce cooling capacity:

– those that act directly on the compressor;

– those that act on the refrigeration circuit.

Very common use

– On–off control: running at full power and then stopping the compressor;

– compressor speed variation;

– the isolation delay of the displaced volume at suction: drawer provided on screw compressors;

– pre-rotation of the vapor at suction: applies to centrifugal compressors by action on the orientation of the velocity of the entering fluid with respect to the blades.

Common use

– On–off control: distribution of the thermal load over several compressors which operate in on–off mode;

– the elimination of cylinders: maintaining the suction valve in the open position.

Uncommon use and/or to be avoided

– The variation of the dead volume of the compressor: delay in the opening of the suction valve;

– HP–LP short circuit: reinjection of part of the vapor which has been compressed in the space under low pressure;

– throttling of the vapor at suction: expansion of the vapor in a valve before entering the compressor.

1.2. Systems driven by thermal energy

1.2.1. *Principle of thermodynamic operation*

The principle of systems driven by thermal energy is to no longer use mechanical energy to produce heat or cold. For this, it is necessary to have a third thermal source at a temperature higher than that of the hot source (at room temperature), which associated with this last source makes it possible to replace the mechanical compressor by a so-called thermal compressor (a device aspirating vapor at low pressure and discharging it at high pressure using only heat). This thermal compressor can be schematized thermodynamically by the association of an engine and a generator (Figure 1.12).

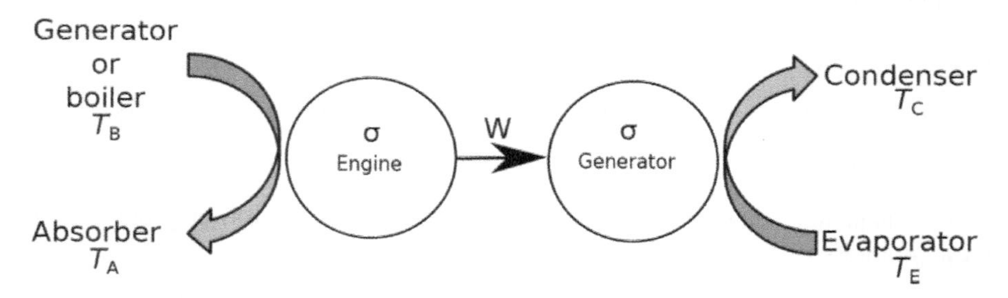

Figure 1.12. *Principle of operation of a machine driven by thermal energy. Example of an absorption machine*

1.2.2. *Absorption chillers*

1.2.2.1. *Operating principle*

Absorption chillers operate by dissolving a vapor in a liquid and its desorption, which respectively releases and absorbs heat. They also use the fact that solubility depends on temperature and pressure. Dissolution achieves the suction phenomenon of the thermal compressor; desorption achieves its discharge.

Thus, these machines necessarily use two fluids: the refrigerant and the liquid solvent (if it is a solid, we refer to adsorption). In most industrial installations, these fluids are ammonia and water, respectively, so this pair will be used in the remainder of this chapter (sorbent or solvent = water; sorbate or refrigerant or solute = ammonia). We could also encounter water as the sorbate and lithium bromide as the sorbent.

Schematically, such an installation comprises (Figure 1.13) the following:

Boiler: the solution receives the quantity of heat Q_B allowing the desorption of the ammonia. The ammonia vapor leaves the boiler at 2. In order to eliminate as much water vapor as possible, which tends to accompany the ammonia, the boiler is equipped with a separator (or rectifier), which consists of a certain number of trays (similar to distillation trays). The weakened solution leaves the boiler at 7. The heat is supplied to the boiler, either by direct heating or by circulation of superheated water or vapor.

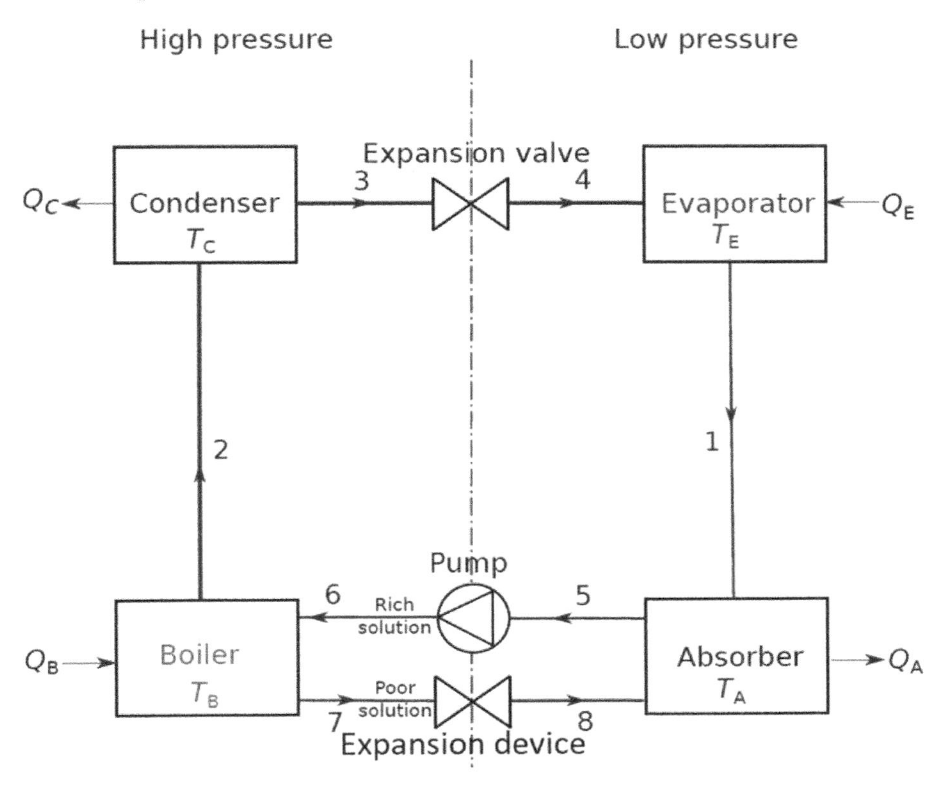

Figure 1.13. *Diagram of an absorption machine*

Condenser: a similar device to compressor refrigerators. The temperature $T_A = T_C$ of the refrigerant in the condenser controls the condensing temperature and pressure, and therefore the pressure in the boiler and the condenser.

Evaporator: at the outlet of the condenser, the high-pressure liquid is throttled through the expansion valve and then evaporates, extracting the heat Q_E from the fluid or the enclosure to be cooled. The temperature T_E and the pressure in the evaporator are controlled by the temperature of the cold source.

Absorber: the ammonia vapor meets the weakened solution (i.e. depleted in ammonia) coming from the boiler there. It dissolves in this solution releasing the heat Q_A that is carried away by a circulation of water, which is generally the same as that which passes through the condenser. The solution thus strengthened (i.e. enriched in ammonia) is sent to the boiler.

The pressure difference between the absorber and the boiler requires:

– the presence of a pump between 5 and 6;

– an expansion device between 7 and 8.

Note that the work done by this pump is significantly lower than that of a compressor. Indeed, in these two cases, it is a question of increasing the pressure of a fluid from P_1 to P_2, but for the pump this fluid is a liquid, whereas for the compressor it is a vapor whose specific volume is much larger than for the liquid. With the specific work being of the form $w = \int vdp$, the specific work of the pump is approximately 1/20 of the specific work of the compressor.

1.2.2.2. *Energy balance*

This study is carried out at the scale of the system considered as closed.

We will denote the below as follows:

Q_B = heat absorbed by the boiler;

Q_E = heat received by the evaporator;

W_t = mechanical energy received by the pump and transmitted to the fluid;

Q_C = heat released by the condenser;

Q_A = heat given up by the absorber.

The first law of thermodynamics applied to the system of Figure 1.13 gives:

$$Q_B + Q_E + W_t = Q_A + Q_C \qquad [1.21]$$

To simplify the reasoning, note that:

– the boiler temperature is T_B;

– the condenser and absorber temperatures are T_M;

– the evaporator temperature is T_E.

So, assuming that all evolutions are reversible, the entropy generation must be zero. Therefore:

$$\frac{Q_B}{T_B}+\frac{Q_E}{T_E}-\frac{Q_C}{T_M}-\frac{Q_A}{T_M}=0 \qquad [1.22]$$

By grouping, $Q_A + Q_C = Q_{AC}$, the following is obtained:

$$\frac{Q_B}{T_B}+\frac{Q_E}{T_E}=\frac{Q_{AC}}{T_M} \qquad [1.23]$$

By dismissing the energy exchanged at the level of the pump, the equation of the first law becomes:

$$Q_B + Q_E = Q_{AC} \qquad [1.24]$$

And therefore

$$Q_B\left(\frac{1}{T_B}-\frac{1}{T_M}\right)+Q_E\left(\frac{1}{T_E}-\frac{1}{T_M}\right)=0 \qquad [1.25]$$

The energy efficiency ratio for a reversible evolution, always defined as the ratio of the thermal energy supplied to the cold source Q_E to the energy supplied to the system ($Q_B + W_t$, where W_t is dismissed), is:

$$\varepsilon=\frac{Q_E}{Q_B}=\frac{T_E}{T_M-T_E}\frac{T_B-T_M}{T_B}=\varepsilon_{Carnot}\eta_{Carnot} \qquad [1.26]$$

which is written as the product of the efficiencies of a Carnot engine operating between the sources T_B and T_M:

$$\eta_{Carnot}=\frac{T_B-T_M}{T_B} \qquad [1.27]$$

and a Carnot generator producing cold at T_E from a source at T_M

$$\varepsilon=\frac{T_E}{T_M-T_E} \qquad [1.28]$$

As for the COP, it is expressed as the ratio of:

$$COP=\frac{Q_{AC}}{Q_B} \qquad [1.29]$$

if heat is recovered from the condenser and absorber.

1.2.2.3. *Study from diagrams*

1.2.2.3.1. Oldham diagram: theoretical cycle of the solution

This is the most widely used and practical diagram for a solution cycle study. It gives the concentration of the sorbate (refrigerant or solute) in the solution as a function of temperature and pressure. It is a diagram ($Log\ P - {}^{1}/_{T}\ or\ P - \theta$) parameterized in mass fraction (Figure 1.14).

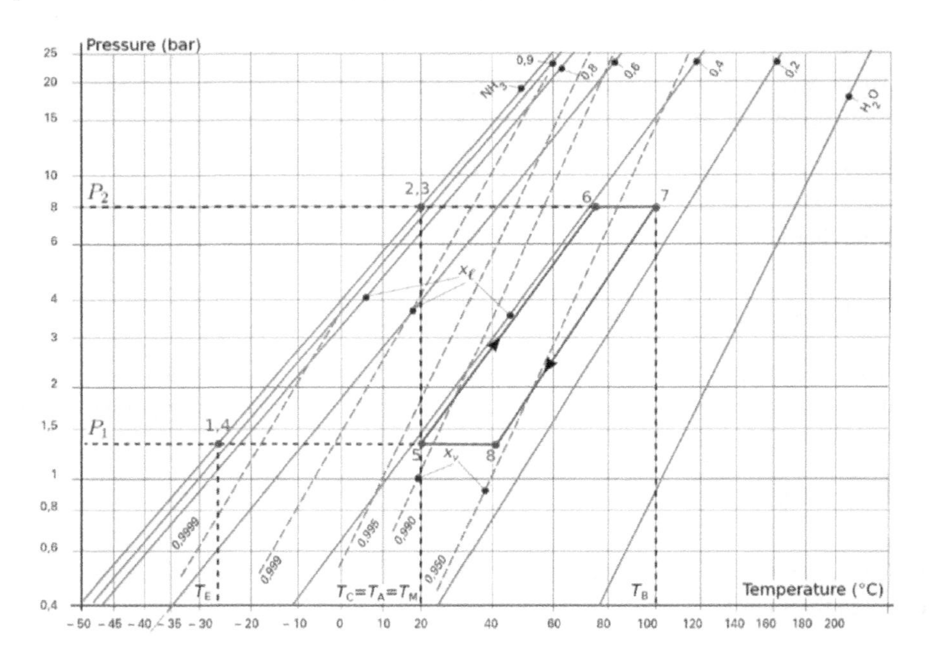

Figure 1.14. *Oldham diagram of the ammonia–water pair. Diagram inspired by Duminil (2002)*

Figure 1.14 includes the plot, in solid lines, of the curves corresponding to the bubble points of five liquid ammonia/water mixtures, whose mass fraction of ammonia is specified. On the left, the diagram includes the curve corresponding to the vapor pressures of pure ammonia and, on the right, the curve of the vapor pressures of pure water. On this diagram, the curves of the dew points of five gaseous mixtures are also plotted, the mass fraction of ammonia of which are indicated, although their usefulness is clearly less. The intersection of the straight line corresponding to the liquid and that corresponding to the vapor gives, for the physical conditions of the point of intersection, the concomitant ammonia fraction of the two phases. They are generally very different, with the gaseous phase being much richer in ammonia (that is more volatile), than the liquid phase.

For the installation described above, the evolution of the solution can be followed on such a diagram (Figure 1.14):

– First, determine the pressure levels P_1 and P_2. These pressures correspond to the saturation vapor pressure of the solute in the evaporator and the condenser respectively. These pressures are obtained by taking the intersection of the isotherms T_M and T_E (it is assumed in this reasoning that the exchanges are ideal) with the line of constant composition $x = 1$ (refrigerant alone). Points 1–4 of the Oldham diagram correspond respectively to points 1–4 of the diagram (Figure 1.13).

– The state of the depleted solution leaving the boiler is given by point 5 of the diagram corresponding to the intersection between the isobar P_2 and the isotherm T_B.

– The content of the solution is obtained before its entry 8 into the absorber, noting that the solution undergoes an expansion at a constant composition between 7 and 8 up to the pressure P_1. Note that expansion involves a drop in temperature due to partial vaporization of the solution. The absorber therefore receives a mixture of liquid solution and vapor (point 8).

– The temperature of the absorber T_A imposes the value of the concentration of solute (refrigerant) at outlet 5 of the absorber. This temperature being at P_1 lower than T_B, there is enrichment of the solution. Point 5 of the diagram gives the value of the concentration x_5.

– Finally, constant pumping will slightly increase the temperature of the solution, as well as its pressure up to P_2. Note that if we assume isentropic compression, the pumping energy is negligible, the liquid will be subcooled and therefore cannot be represented on the Oldham diagram, which only shows liquid–vapor equilibrium states. On entering the boiler, the liquid will therefore first be heated to the temperature T_6. Desorption will then begin at point 6 thus defined.

1.2.2.3.2. Merkel diagram

This diagram allows a complete study of the absorption system because, in addition to the information given by the Oldham diagram, it provides information, which is fundamental for the calculations, on the enthalpy of the solution and the vapor of the solute.

It is a $h - x$ diagram parameterized in pressure and temperature for the solution and in pressure for the vapor mixture. It also makes it possible to know the fraction of the vapor mixture in equilibrium with the solution. It can be seen that, except for very low fractions, the solute content (sorbate or refrigerant) of the vapor is always much higher than that of the solution.

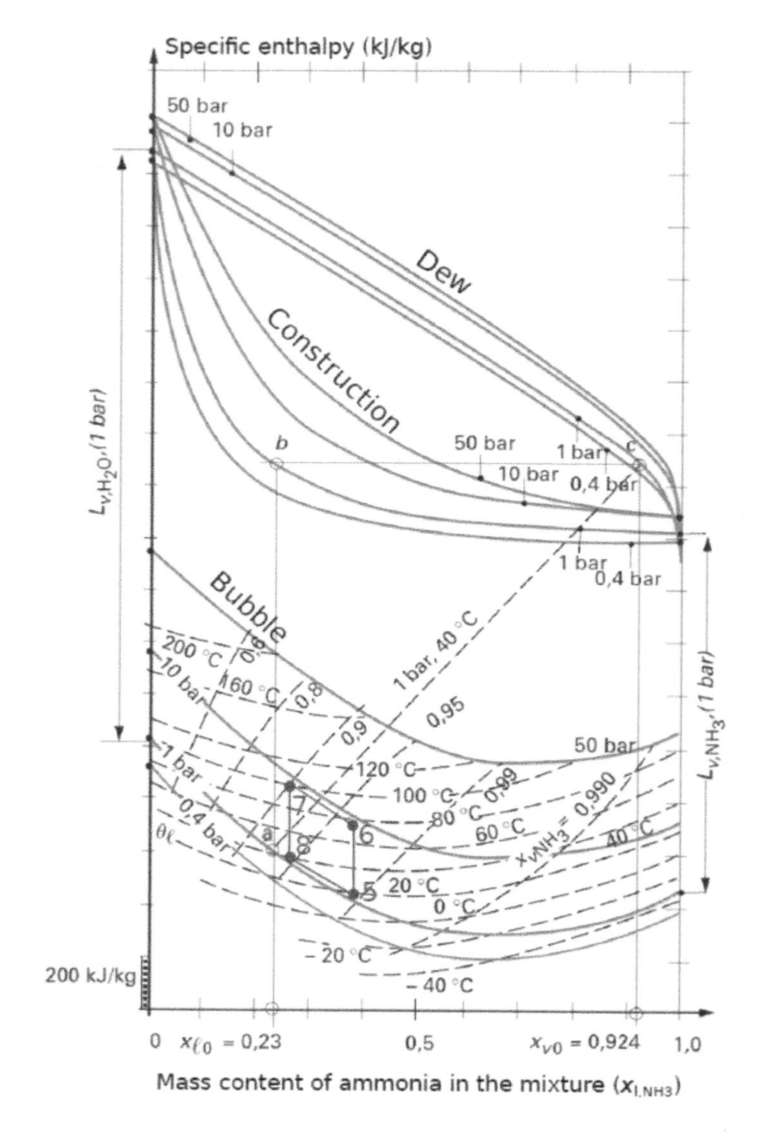

Figure 1.15. *Merkel diagram of the ammonia–water pair. Diagram inspired by Duminil (2002)*

This diagram includes the following curves:

– saddle-shaped isobaric bubble curves: the domain of subcooled liquid mixtures is below each of them for the pressure considered;

– isobaric dew curves: the domain of gaseous mixtures is above each of them for this pressure;

– construction isobaric curves which, like the previous ones, do not have a physical meaning but which are used to find the concomitant equilibrium compositions of the liquid and vapor phases present;

– isotherms of liquid mixtures that are also saddle-shaped;

– sometimes, the curves of equal mass fraction of refrigerant in the gaseous phase, here $x_{v,NH3}$.

On each of the axes, the bubble and dew curves are separated by a segment which represents the enthalpy of vaporization of the component considered. Thus, for a pressure of 1 bar, for example, the segment which separates, on the water side, the extremities of the bubble and dew curves, $L_{v,H2O(1\ bar)}$, corresponds to the enthalpy of vaporization of water under this pressure. On the ammonia side, the segment $L_{v,NH3(1\ bar)}$ represents the enthalpy of vaporization of this fluid under the same pressure.

1.2.2.4. *Thermodynamic overview*

This study is carried out at the scale of the components on which the first law is applied (open system).

G is the solute mass flow rate. The following heat balance is made for the unit mass of solute, which passes through the condenser. The total balance will be obtained by multiplying all of the results by G.

f is the mass of enriched solution that enters the boiler when the unit mass of solute leaves it (Figure 1.13). The mass of depleted solution leaving the boiler will then be $f - 1$. If x_r is the mass fraction of the enriched solution and x_p that of the depleted solution, then:

$$f x_r - (f - 1)x_p = 1 \leftrightarrow f = \frac{1 - x_p}{x_r - x_p} \quad [1.30]$$

Next is an overview, component by component.

– Evaporator overview

$$q_E = h_1 - h_4 \quad [1.31]$$

where:

– h_4 is the specific enthalpy of the refrigerant (solute) in the vapor state under the pressure P_1;

– h_1 is the specific enthalpy of the liquid refrigerant (solute) under P_1.

Specific enthalpy values can be read on the diagram of pure solute (Mollier diagram of ammonia, for example). Note that in fact, the fluid is not the pure solute (see the Merkel diagram), but a mixture of solute (refrigerant or sorbate) and solvent (sorbent). This feature will not be taken into account.

– Boiler overview

Per unit mass of solute leaving the boiler, the enthalpy balance is as follows:

Inlet:

– heat q_B;

– sensible heat $f\ h_6$ due to the entry of the enriched solution.

Outlet:

– sensible heat $(f - 1)h_7$ due to the exit of the depleted solution;

– sensible heat h_2 due to the exit of the solute vapor (refrigerant).

So:

$$q_B = h_2 + (f - 1)h_7 - fh_6 \quad [1.32]$$

where:

– h_2 is the specific enthalpy of the solute vapor under the saturation vapor pressure P_2 from the Merkel diagram, x = 100% or the Mollier diagram of the solute;

– h_7 is the specific enthalpy of the depleted solution, under the pressure P_2, from the Merkel diagram;

– h_6 is the specific enthalpy of the enriched solution, under pressure P_2.

Note that this reasoning obviously requires an origin of the enthalpies common to the solute vapor and to the solution.

– Absorber overview

By analogous reasoning, the following can be written as:

$$q_A = -h_1 - (f-1)h_8 + fh_5 \qquad [1.33]$$

where:

– q_A is in algebraic value;

– h_8 is the specific enthalpy of the depleted solution under the pressure P_1 from the Merkel diagram;

– h_5 is the specific enthalpy of the enriched solution under pressure P_1 from the Merkel diagram.

– Condenser overview

$$q_C = h_3 - h_2 \qquad [1.34]$$

where h_3 is the specific enthalpy of the liquid solute under saturated vapor pressure P_2.

– Pump overview

$$w_t = f(h_6 - h_5) \qquad [1.35]$$

(if the compression is isentropic, the pumping energy is negligible and point 4 coincides with point 5)

Considering these various relationships, the energy efficiency ratio is:

$$\varepsilon = \frac{q_E}{q_B + w_t} = \frac{h_1 - h_4}{h_2 + (f-1)h_7 - fh_6 + f(h_6 - h_5)} = \frac{h_1 - h_4}{h_2 + (f-1)h_7 - fh_5} \qquad [1.36]$$

It should be noted that in practice, this energy efficiency ratio is approximately 0.2 times that which is obtained by the theoretical formula (reversible). Note that ε is improved by allowing a heat exchange between the hot depleted solution which leaves the boiler and the cold enriched solution which leaves the absorber. Such an exchanger will always be present in absorption installations.

The COP is written as (if we recover heat on the condenser and the absorber):

$$COP = \frac{q_C+q_A}{q_B+w_t} = \frac{h_3-h_2-h_1-(f-1)h_8+fh_5}{h_2+(f-1)h_7-fh_6+f(h_6-h_5)} = \frac{h_3-h_2-h_1-(f-1)h_8+fh_5}{h_2+(f-1)h_7-fh_5} \quad [1.37]$$

1.2.2.5. *Operating limits*

The operating limits of the system are described below and in Figure 1.16. This text is inspired by Duminil (2002).

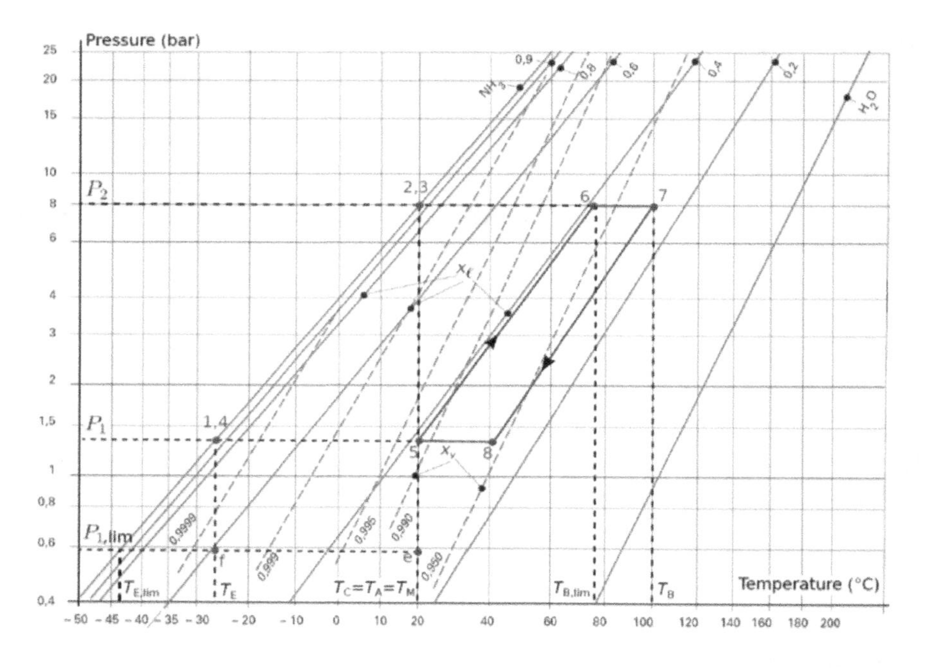

Figure 1.16. *Oldham diagram (enthalpy-concentration) of the ammonia–water pair: operating limits. Diagram inspired by Duminil (2002)*

Theoretical lower limit of the motive heat source temperature: $\boldsymbol{T_B}$

T_M and T_m are fixed and so P_1, P_2, x_r and points 5 and 6 are also fixed.

By decreasing T_B, point 7 is brought closer to point 6; The machine theoretically stops working when $x_r = x_p$ (a real absorption chiller stops working long before this condition). An important practical deduction is drawn from this, which is also valid for real systems: the thermal energy that supplies the generator of an absorption system must be supplied at a sufficiently high temperature. This consideration takes on its full meaning when considering producing cold using industrial waste heat, solar heat, etc.

Theoretical lower limit of the cold source temperature: $\boldsymbol{T_m}$

T_B and T_M are fixed and so P_2, point 7, x_p and point e are also fixed.

If T_m decreases, the pressure P_1 decreases, and point 5 "descends" along the isotherm T_M. The solution leaving the absorber (point 5) is gradually weaker in the refrigerant. Theoretically, the absorption system stops working when points 5 and e coincide ($x_r = x_p$). The lower limit of the temperature of the cold source, $T_{m,lim}$, is directly determinable (here about -43°C). This theoretical limit is lowered by raising the temperature of the generator T_B. Hence, the important consideration, also valid for real systems: the temperature of the thermal energy supply source must increase as the cold is produced at a lower temperature.

Theoretical upper limit of cold sink temperature: $\boldsymbol{T_M}$

T_B and T_m are fixed and so P_1 is also fixed.

When T_M rises, leading to an increase in pressure P_2 in the generator–condenser assembly, point 5 moves to the right, point 7 rises along the isotherm T_B, the cycle of solutions grows longer and thinner and $x_r \rightarrow x_p$. This limit can only be raised by increasing T_B. From where the practical consequence, true for the real systems as for the perfect ones: the increase, even minor, in the temperature of the heat sink has very harmful effects on the operation of the system with absorption; the temperature of this sink should be kept as low as possible.

1.2.2.6. *Different architectures of the absorption system*

In addition to the choice of fluid, an improvement in the performance of the absorption system can be obtained by modifying the architecture of the cycle.

From the basic single-acting cycle (Figure 1.13), several modifications can be made: adding additional components, coupling certain components, replacing components with other elements.

The objective of a modification of the architecture is to improve the performance of the cycle to obtain better efficiency, higher operating power or more satisfactory output variables (e.g. lower evaporation temperature or a higher temperature difference between the different levels).

The Generator Absorber heat eXchanger (GAX) cycle consists of "reuniting" the generator and the absorber. This "contacting" makes it possible to recover the heat from the absorber to heat the solution in the generator. Thus, the heat transferred to the generator is lower. Most GAX systems use the ammonia–water pair and almost all GAX systems are used for refrigeration. There are two types: simple GAX cycle and branched GAX cycle. The value of the energy efficiency ratio can be increased by 20–30% compared to that of the conventional cycle for the same operating conditions. Compared to the simple GAX cycle, the branched GAX cycle has an additional solution pump between the generator and the absorber in order to increase the mass flow in the absorber.

The multi-effect cycles also allow an increase in the efficiency of the installation. In particular, in a double-acting cycle, two generators are present: the high-pressure generator (HPG) and the low-pressure generator (LPG). The refrigerant is first desorbed in the HPG. Then, the vapor produced is used to heat the solution in the LPG. The vapor thus desorbed from the solution in the LPG is then condensed in the condenser with the vapor from the HPG. The solution is therefore better desorbed with two generators, and the efficiency of the desorption process increases compared to the single-effect cycle. Triple-effect cycle devices also exist on the same principle with three generators.

There are also multi-stage cycles such as the two-stage absorption cycle. It is a combination of two single-stage absorption cycles by coupling the evaporator of the first cycle and the condenser of the second cycle.

Finally, adding an ejector to the system is also a solution to increase the energy efficiency ratio or the COP. The ejector is added between the rectifier (component placed after the generator to ensure quality desorption) and the condenser. The high-pressure primary ammonia vapor from the rectifier then enters the ejector and draws in the low-pressure vapor. The mixture is then introduced into the condenser. The interest is to increase the flow rate in the evaporator and therefore the cooling power produced for a condensing pressure lower than that of desorption.

Many other solutions are proposed in the literature.

1.2.2.7. *Application example*

We propose to study a 100 kW single-effect chiller operating with the water–lithium bromide pair. The pressure drops are neglected, and the temperatures in the evaporator and the condenser are uniform and equal to 5°C and 30°C, respectively. The temperature of the end of regeneration is taken as equal to 75°C, and the temperature of the end of absorption is equal to 30°C. A heat exchanger on the solution is used with an efficiency of 0.78. The pumping energy is neglected, as well as the thermal effect of the pump.

As shown in Figure 1.17, point 1 is at the inlet of the thermochemical compressor and point 2 is at its outlet. Points 3 and 4 are respectively upstream and downstream of the expansion valve. Points 10 and 6 are respectively located at the inlet and outlet of the intermediate exchanger on the strong solution side, while points 7 (inlet) and 9 (outlet) are found at the terminals of the exchanger on the weak solution side. Point 6 is more precisely located at the inlet of the boiler and is saturated. Finally, point 5 is located at the pump inlet (strong solution) and point 8 is located at the expansion device outlet (weak solution), i.e. at the inlet of the absorber (it is saturated).

We will seek to calculate the flow rates of the strong and weak solutions, as well as the energy efficiency ratio with and without intermediate exchanger.

Table 1.2 shows the properties of fluids at different points in the cycle.

	P (bar)	*T* (°C)	*h* (kJ/kg)	Concentration of BrLi
1	0.0087	5	2510	0
2	0.04247	30	2635	0
3	0.04247	30	125.7	0
4	0.0087	5	125.7	0
5	0.0087	30	66.35	0.52
6	0.04247	57.85	113.3	0.52
7	0.04247	75	188.6	0.61
8	0.0087	48.85	133.5	0.61
9	0.04247	39.9	133.5	0.61
10	0.04247	30	66.35	0.52

Table 1.2. *Thermodynamic properties of fluids at different points in the cycle*

To complete the table, the Odham and Merckel diagrams for the water–lithium bromide pair are used. In addition, the temperature of point 9 was obtained from the efficiency of the exchanger:

$$\begin{aligned} \varepsilon &= \frac{T_7 - T_9}{T_7 - T_{10}} = 0.78 \\ T_9 &= T_7 - 0.51 \cdot (T_7 - T_{10}) \\ &= 39.9\ °\text{C} \\ &= 312.9\ \text{K} \end{aligned}$$

The pumping energy and the thermal effect of the pump are negligible. Therefore, $h_5 = h_{10}$. The expansions in the expansion valve and the expansion device are isenthalpic. Thereby, $h_3 = h_4$ and $h_8 = h_9$.

To calculate the enthalpy of point 6, it is necessary to carry out an energy balance on the exchanger:

$$\begin{aligned} 0 &= \dot{m}_6 \cdot (h_6 - h_{10}) + \dot{m}_7 \cdot (h_9 - h_7) \\ h_6 &= -\frac{\dot{m}_7}{\dot{m}_6} \cdot (h_9 - h_7) + h_{10} \end{aligned}$$

However, it is known that the lithium bromide mass flow rates at points 6 and 7 are identical. It becomes:

$$\begin{aligned} \dot{m}_{6,\text{LiBr}} &= \dot{m}_{7,\text{LiBr}} \\ x_6 \cdot \dot{m}_6 &= x_7 \cdot \dot{m}_7 \\ \frac{\dot{m}_7}{\dot{m}_6} &= \frac{x_6}{x_7} \end{aligned}$$

with x being the mass fraction of lithium bromide. Thus, the enthalpy of point 6 is expressed by:

$$\begin{aligned} h_6 &= -\frac{x_6}{x_7} \cdot (h_9 - h_7) + h_{10} \\ &= -\frac{0{,}52}{0{,}61} \times (133.5 - 188.6) + 66.35 \\ &= 113.3\ \text{kJ/kg} \end{aligned}$$

Once the table is completed, the mass flow rates of the fluids can be calculated. For this, it is first necessary to carry out an energy balance on the boiler followed by a balance on the evaporator. With G being the mass flow rate of water in the evaporator, the balance on the boiler gives:

$$\dot{m}_6 = \dot{m}_7 + G = \dot{m}_6 \cdot \frac{x_6}{x_7} + G = \frac{x_7}{x_7 - x_6} \cdot G$$

but the balance on the evaporator is written as:

$$G = \frac{\dot{Q}_{\text{cold}}}{h_1 - h_4} = \frac{100}{2510 - 125.7} = 41.9 \text{ g/s}$$

Hence, the following flow values:

$\dot{m}_6 = \frac{x_7}{x_7 - x_6} \cdot G = 285 \text{ g/s}$ corresponding to the flow rate of the strong solution;

$\dot{m}_7 = \dot{m}_6 \cdot \frac{x_6}{x_7} = 243 \text{ g/s}$ corresponding to the flow rate of the weak solution.

We can now calculate the energy efficiency ratio

$$\varepsilon = \frac{\dot{Q}_{\text{cold}}}{\dot{Q}_{\text{generator}}} = \frac{100}{124} = 0.8$$

with

$$\begin{aligned}\dot{Q}_{\text{generator}} &= G \cdot h_2 + \dot{m}_7 \cdot h_7 - \dot{m}_6 \cdot h_6 \\ &= 0.0419 \times 2635 + 0.243 \times 188.6 - 0.285 \times 113.3 \\ &= 124 \text{ kW}\end{aligned}$$

Note that without an intermediate exchanger, the energy efficiency ratio would be:

$$\varepsilon = \frac{\dot{Q}_{\text{cold}}}{\dot{Q}_{\text{generator,SE}}} = \frac{100}{137.5} = 0.73$$

with

$$\dot{Q}_{\text{generator,SE}} = \dot{m}_7 \cdot h_7 + \dot{m}_2 \cdot h_2 - \dot{m}_6 \cdot h_{10} = 137.5 \text{ kW}$$

The intermediate exchanger saves about 10% of the energy efficiency ratio.

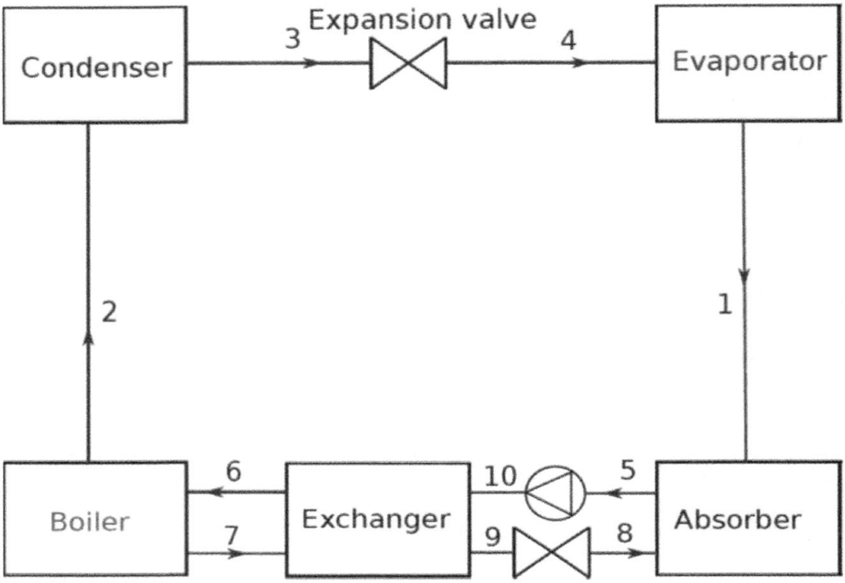

Figure 1.17. *Diagram of the installation of an absorption chiller with intermediate heat recovery exchanger*

1.2.3. *Ejection machines*

1.2.3.1. *Operating principle*

Ejection machines are a second example of a three-heat-source system driven by thermal energy. In this case, the thermal compressor essentially consists of an ejector. The ejector draws the refrigerant in the evaporator, compresses it and pushes it back into the condenser. Fluid compression is achieved by momentum transfer in the ejector. This momentum is provided by the expansion of high-pressure vapor taken from a boiler (Figure 1.18).

It is this transfer of momentum that plays the role of the dissolution phenomenon in absorption machines. The advantage here is to use only one working fluid, unlike the absorbate–solvent pair of absorption machines. The ejection machine thus comprises a high-pressure boiler supplied with liquid by a pump, the ejector, the evaporator supplied with a two-phase fluid by an expansion valve and finally a condenser at the outlet of the ejector to complete the cycle. It is a simple and rustic system that does not use any mechanical element except the pump.

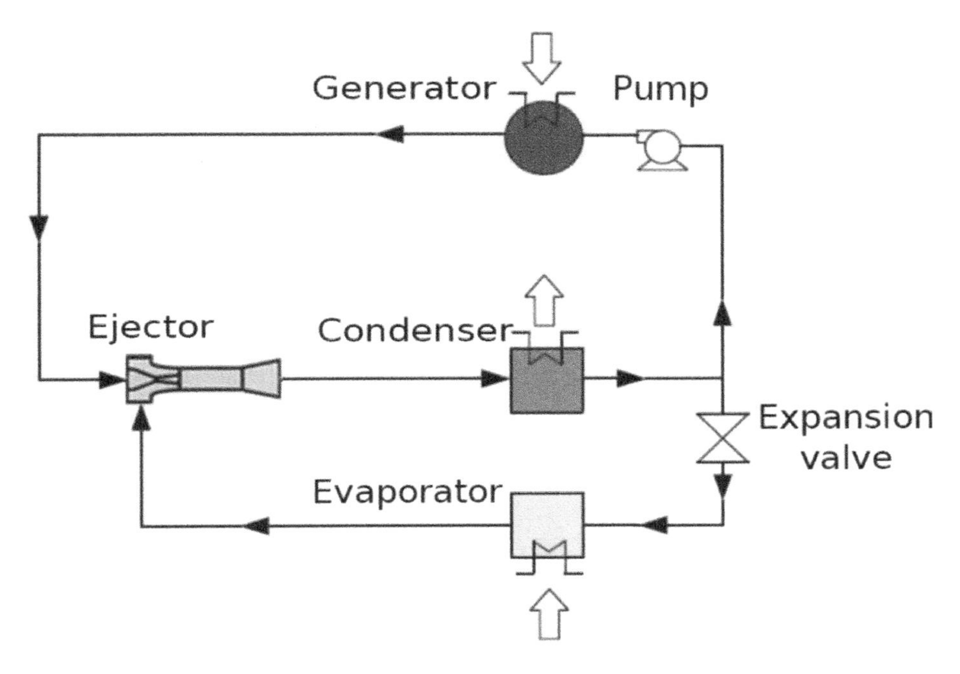

Figure 1.18. *Operating diagram of an ejection refrigerating system (Bouzrara 2018)*

The cold produced depends on the operation of the ejector, which is governed by the three pressures which prevail at its orifices. These pressures are related to the three temperatures of the sources and to the nature of the working fluid.

Operation of ejectors

The ejectors consist of two coaxial nozzles linked together by a suction duct (Figure 1.19). The primary fluid, at high pressure, coming from the boiler, enters the ejector through the so-called motive nozzle where it reaches the speed of sound at the neck (plane i), also called the throat. The vapor coming from the evaporator, which constitutes the secondary fluid, enters an annular space around the motive nozzle called the suction or stilling chamber. At the outlet of the motive nozzle diffuser (plane ii), the supersonic velocity of the primary fluid is such that the pressure is lower than the pressure of the secondary fluid, which allows it to be sucked into the mixing chamber. In this chamber, an exchange of momentum takes place. Thus, the secondary fluid is accelerated and mixed with the primary fluid. At the neck of the mixing chamber, a shock wave takes place (plane v), which compresses the mixture and reduces the velocity below the speed of sound. In the subsonic diffuser, the compression of the fluid continues up to the condensing pressure.

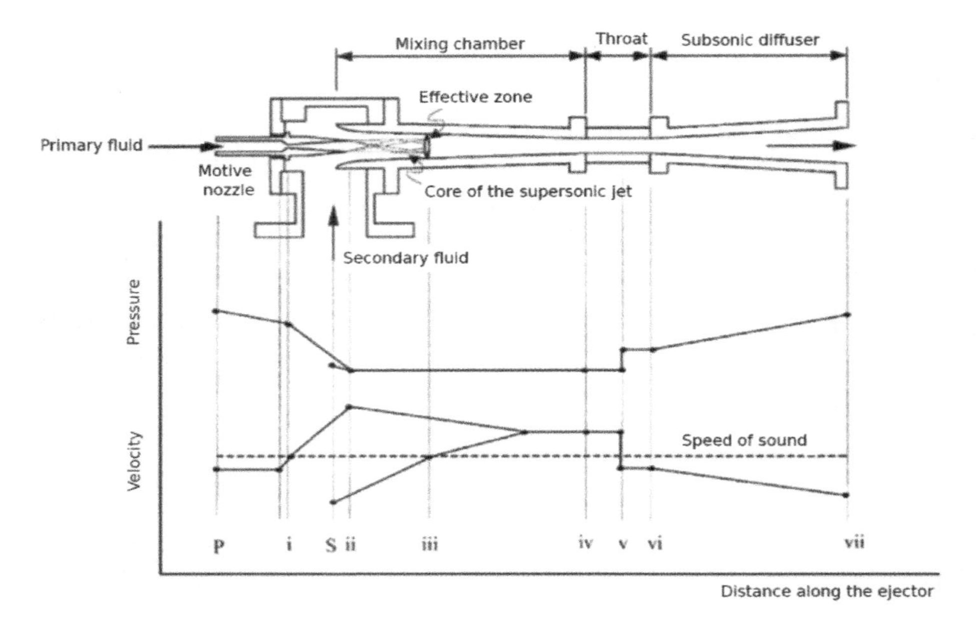

Figure 1.19. *Schematic view of an ejector and changes in pressure and speed as a function of position (Chunnanond and Aphornratana 2004)*

1.2.3.2. *Energy balance*

With the notations proposed in section 1.2.2.2, the energy balance of an ejection refrigerating machine is written as:

$$Q_B + Q_E + W_t = Q_C \quad [1.38]$$

And for a reversible machine:

$$\frac{Q_B}{T_B} + \frac{Q_E}{T_E} - \frac{Q_C}{T_M} = 0 \quad [1.39]$$

Neglecting the energy exchanged in the pump, the equation of the first law becomes:

$$Q_B + Q_E = Q_C \quad [1.40]$$

And the expression of the energy efficiency ratio:

$$\varepsilon = \frac{Q_E}{Q_B} = \frac{T_E}{T_M - T_E}\frac{T_B - T_M}{T_B} = \varepsilon_{\text{Carnot}}\eta_{Carnot} \quad [1.41]$$

These expressions could have been directly formulated given the fact that they only depend on the use of the three sources.

The COP is expressed by the ratio between

$$COP = \frac{Q_C}{Q_B} \quad [1.42]$$

1.2.3.3. *Study from diagrams*

The cycle followed by the fluid in the machine is shown in Figure 1.20. The numerical references correspond to the ends of the exchangers, i.e. for the boiler 2-3, the condenser 4-1, the evaporator 5-6, the pump 1-2 and the ejector 3-6-4. The energy efficiency ratio is written as:

$$\varepsilon = \frac{Q_E}{Q_B} = \frac{\dot{m}_s\,(h_6 - h_5)}{\dot{m}_p\,(h_3 - h_2)} \quad [1.43]$$

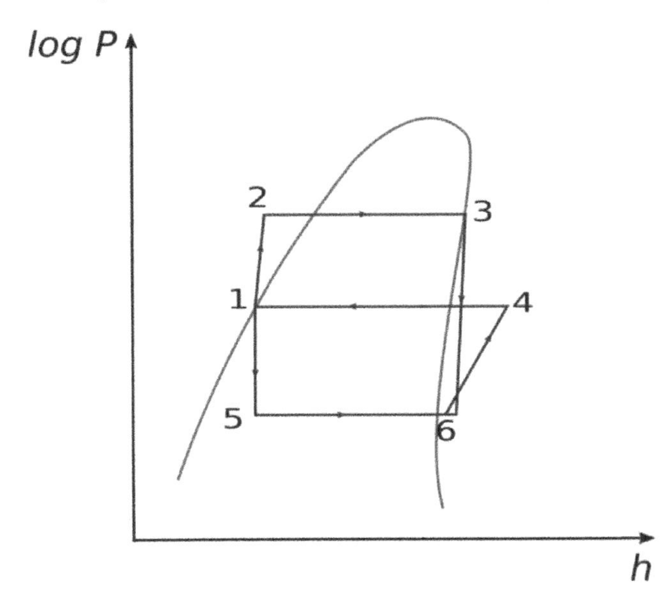

Figure 1.20. *Enthalpy diagram of an ejection refrigerating system*

where $\dot{m}_p$ and $\dot{m}_s$ are the primary (set by the pump) and secondary (sucked in by the ejector) flows respectively. By introducing the entrainment ratio $\mu = {}^{\dot{m}_s}/_{\dot{m}_p}$,

the ratio between the secondary and primary flow rates is:

$$\varepsilon = \mu \frac{(h_6 - h_5)}{(h_3 - h_2)} \qquad [1.44]$$

The enthalpy differences vary relatively little with the operating conditions, so the performance of the system is closely dependent on the entrainment ratio of the ejector, which is its main characteristic.

1.2.3.4. *Characterization of ejectors*

For a given refrigerant and a fixed ejector geometry, in most cases, the flow is sonic in the primary nozzle. In this case, the engine flow rate depends only on the conditions (temperature and pressure) at the boiler outlet. It can be calculated knowing the speed of sound at the throat and the cross-section. The entrainment ratio depends on the three pressures at the ends of the ejector. Some authors suggest using two pressure ratios instead: $\pi_{s0,s2} = P_E/P_B$ (reduced low pressure) and $\pi_{s1,s2} = P_C/P_B$ (reduced counter-pressure).

In nominal operation, the primary flow is supersonic at the outlet of the primary diffuser and the secondary flow is sonic in a narrow section at the periphery of the primary jet called the effective zone (Figure 1.21). For a given value of the reduced low pressure, these conditions are achieved from a minimum reduced counter-pressure, up to the optimum pressure corresponding to the limit of the extinction of the shock wave in the diffuser (Figure 1.22). Because of the blocking of flows due to their criticality, the entrainment ratio depends only on the reduced low pressure and exhibits a plateau depending on the counter-pressure. Beyond the optimum, the secondary flow decreases with increasing counter-pressure, until it reverses for maximum counter-pressure. In the space $\mu - \ln(\pi_{s0,s2}) - \ln(\pi_{s1,s2})$, the doubly sonic operating points are in a plane (Figure 1.22). For a given reduced counter-pressure, the entrainment ratio increases with reduced low pressure.

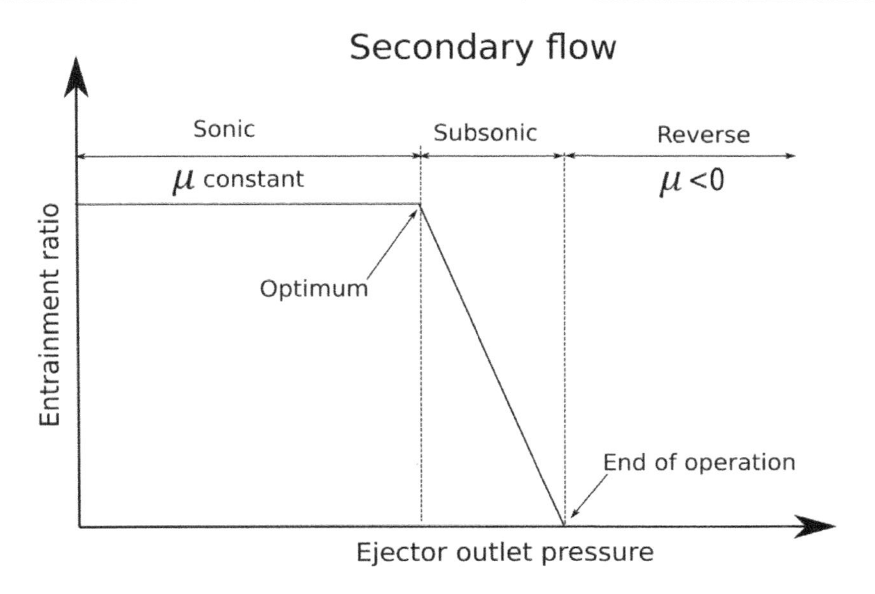

Figure 1.21. *Influence of the ejector outlet pressure (counter-pressure) on the entrainment ratio at a constant reduced low pressure*

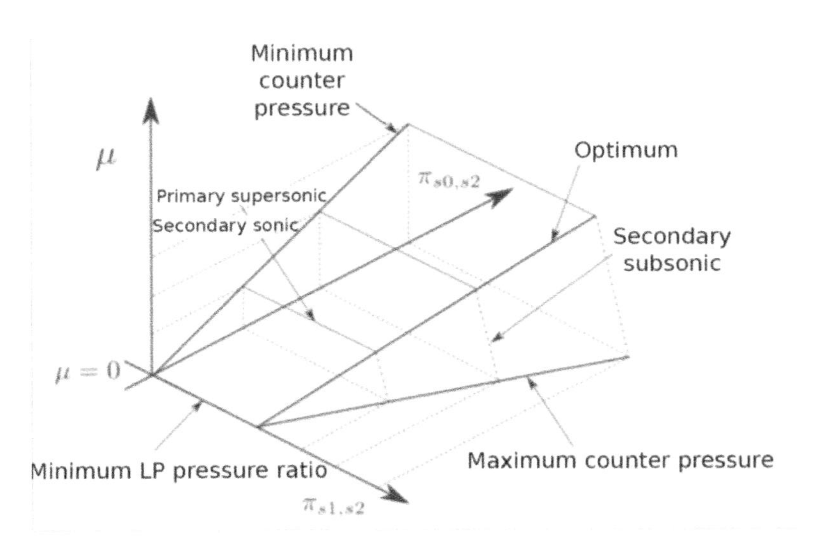

Figure 1.22. *Characteristic field of an ejector. Figure inspired by Zegenhagen and Ziegler (2015)*

1.3. References

Bouziane, A. (2014). Simulation et étude expérimentale d'une machine frigorifique au CO_2 transcritique munie d'un éjecteur. PhD Thesis, INSA Lyon.

Bouzrara, A. (2018). Etude expérimentale des éjecteurs : application à la récupération de l'énergie de détente des machines frigorifiques au CO_2. PhD Thesis, INSA Lyon.

Chunnanond, K., and Aphornratana, S. (2004). Ejectors: Applications in refrigeration technology. *Renew. Sustain. Energy Rev.*, 8(2), 129–155. doi: 10.1016/j.rser.2003.10.001.

Duminil, M. (2002). Machines thermofrigorifiques – Systèmes à éjection et à absorption. *Techniques de l'ingénieur*, be9735.

Guangming, C., Xiaoxiao, X., Shuang, L., Lixia, L., Liming, T. (2010). An experimental and theoretical study of a CO_2 ejector. *International Journal of Refrigeration*, 33, 915–921.

Zegenhagen, M.T. and Ziegler, F. (2015). Experimental investigation of the characteristics of a jet-ejector and a jet-ejector cooling system operating with R134a as a refrigerant. *International Journal of Refrigeration*, 56, 173–185.

2

Entropy and Exergy Analyses Applied to Reverse Cycles

Jocelyn BONJOUR and Rémi REVELLIN
CETHIL, INSA Lyon, Villeurbanne, France

2.1. Definition of the study system and objectives

The first chapter of this book focuses on a rather practical and technical approach to describe the state of the art relating to the most common reverse cycle engines. It also underlines the necessary distinction between ideal operation ("reversible", in the thermodynamic sense of the term) and real functioning (Chapter 1, section 1.1.1). This results in a real challenge for the energy engineer to assess the deviation from reversibility as a performance indicator for the systems it is studying. This is precisely what entropy analysis and exergy analysis, which derives from a combination of energy and entropy analyses, allow. Let us also insist on the fact that this chapter focuses on the *analysis* of reverse cycle engines, not on their *design*. It is therefore a question of understanding the operation of an engine from the study of its components and their relationships, and not of choosing components, or sizing them, to meet specifications.

This is why entropy and energy analyses can be carried out at the scale of the complete system (i.e. of the reverse cycle engine), but an analysis on the scale of the components, or even of the physical phenomena present, allows us to better identify ways to improve the systems. For example, the distribution of irreversibility (i.e. of

For a color version of all the figures in this chapter, see www.iste.co.uk/bonjour/refrigerators. zip.

Refrigerators, Heat Pumps and Reverse Cycle Engines,
coordinated by Jocelyn BONJOUR.

the destruction of exergy) between the components makes it possible to identify the component which must be improved as a priority. In this chapter, the required thermodynamic methods and notions will be introduced as they are used, some being close to but complementary to those developed in Chapter 1. It is also assumed that the reader is familiar with the main thermodynamic variables (state functions: in particular enthalpy, entropy, internal energy, which will often be used as specific quantities, i.e. per unit mass) and the intensive or extensive quantities essential in thermodynamics, such as pressure, temperature, volume or specific volume, etc.

Generally speaking, a reverse cycle engine can usually be represented schematically as a closed thermodynamic system (Figure 2.1) which makes it possible to extract thermal power (this is cold generation, refrigerating power $\dot{Q}_f$) from a cold source by receiving mechanical power ($\dot{W}$) or thermal power $\dot{Q}_m$ (one or the other (or even both) constitutes the motive power of the system), which leads to the release of heat (heating power $\dot{Q}_f$) to a hot source. The engine is said to be a two-heat-source system if only the hot and cold sources exchange heat with it, i.e. if it is a mechanical power which is the motive power of the system. It is said to be a three-heat-source system if its motive power is the thermal power $\dot{Q}_m$, exchanged with a third source at another temperature. It is also common for this system to exchange heat with its environment in a diffuse and poorly controlled manner: these are heat gains and/or rejections (thermal power $\dot{Q}_p$), which are not necessarily negligible, although an abundant part of the available literature admits, often implicitly, the adiabaticity of the system. These engines are called refrigerators or refrigerating systems if the service rendered to the user consists of cold generation, and heat pumps if the service rendered consists of heat generation from heat rejection. A fluid (or several fluids) generally travels through a circuit made up of several components. For example, in the case of the dominant technology of cold generation by mechanical vapor compression (Chapter 1), four essential components are generally retained (Figure 2.2):

– the evaporator, which allows heat exchange with the cold source ($\dot{Q}_{evap} = \dot{Q}_f$) as the refrigerant evaporates;

– the condenser, which allows heat exchange with the hot source ($\dot{Q}_{cond} = \dot{Q}_c$) as the refrigerant condenses;

– the compressor which compresses the refrigerant;

– the expansion valve which ensures the expansion of the refrigerant.

The heat gains and rejections can be located on the circulation lines or on the component materials.

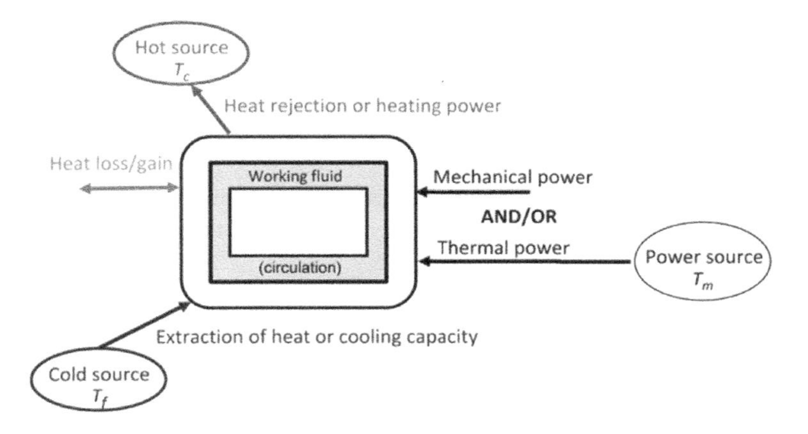

Figure 2.1. *Conceptual diagram of a reverse cycle engine*

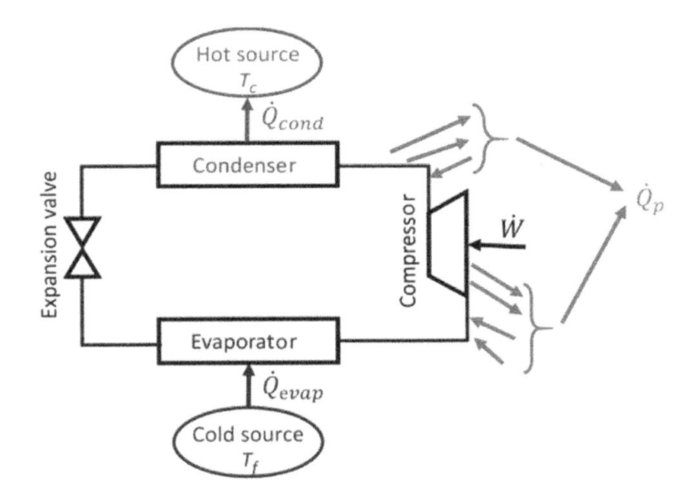

Figure 2.2. *Circuit of a vapor compression reverse cycle engine*

In the case of an absorption refrigerator or heat pump (Chapter 1), the following components (Figure 2.3) are retained:

– the boiler, allowing the refrigerant to be desorbed (thermal power absorbed $\dot{Q}_m$, extracted from a heat source at the temperature T_m);

– the evaporator, similar to that of compression systems (thermal power $\dot{Q}_{evap}$ exchanged with the hot source at the temperature T_f);

– two expansion devices (known as adiabatic as for vapor compression technology): the expansion valve and the regulating valve;

– the condenser, whose role is the same as for mechanical compression systems (thermal power $\dot{Q}_{cond}$ exchanged with the hot source at the temperature T_c);

– the absorber, which allows the absorption of the refrigerant, an exothermic process leading to heat rejection $\dot{Q}_{abs}$, generally towards the hot source (therefore also at the temperature T_c);

– a pump, consuming a certain mechanical power $\dot{W}$ (a priori modest compared to the motive power of heat $\dot{Q}_m$).

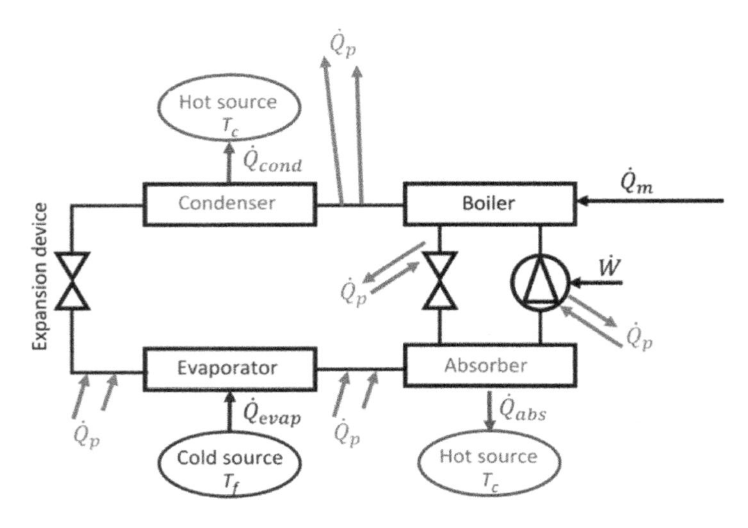

Figure 2.3. *Circuit of an absorption reverse cycle engine*

2.2. Energy analysis

Energy analysis proceeds from the first law of thermodynamics, applied to the complete system or to each component. It is usual to reason in powers and in steady state, or in energy integrated over a certain duration.

The first law of thermodynamics applied to a closed system is written as:

$$\frac{dE}{dt} = \Sigma(\dot{W}_i + \dot{Q}_i) \qquad [2.1]$$

E represents here the total energy of the system, which is reduced to its internal energy in the absence of variation of kinetic energy or potential energy. $\dot{W}_i$ and $\dot{Q}_i$ respectively designate the mechanical and thermal powers exchanged by the system with its external environment. According to the so-called "banker" sign convention, input quantities are positive (heat absorbed, work done on the system, etc.) and output quantities are negative (heat rejected, work done by the system, etc.).

For a flow of matter through an open system, where the effects of kinetic and potential energy are neglected:

$$\frac{dE}{dt} = \Sigma(\dot{W}_i + \dot{Q}_i) + \Sigma(\dot{m}_e h_e - \dot{m}_s h_s) \qquad [2.2]$$

where $\dot{m}_e$ denotes entering specific flow rates, $\dot{m}_s$ denotes exiting specific flow rates, h_e denotes the specific enthalpy of the entering matter flux and h_s denotes the specific enthalpy of the exiting matter flux. Note that a negative sign is assigned to the term carrying the exiting flow: the flow values here are always counted positively.

2.2.1. *Steady-state system-wide analyses*

For a reverse cycle engine such as the one in Figure 2.2, the first law of thermodynamics applied to the complete system (closed system) is written as:

$$\frac{dE}{dt} = \dot{W} + \dot{Q}_{cond} + \dot{Q}_{evap} + \dot{Q}_p \qquad [2.3]$$

In steady state, it is reduced to:

$$0 = \dot{W} + \dot{Q}_{cond} + \dot{Q}_{évap} + \dot{Q}_p \qquad [2.4]$$

In accordance with the sign convention adopted previously, it should be noted that $\dot{W} > 0$ because the refrigerant receives work, $\dot{Q}_{cond} < 0$ because the refrigerant rejects heat towards the hot source and $\dot{Q}_{evap} > 0$ because the refrigerant absorbs heat from the cold source. $\dot{Q}_p$ can be globally positive (more gains than rejections), negative (in the opposite case) or zero (if the components are isolated, or if all the gains and rejections compensate each other).

The steady-state energy analysis leads to the definition of energy performance indicators for the systems studied, which are defined in the following sections.

2.2.1.1. *Two-heat-source heat pumps and refrigerators*

Depending on whether the reverse cycle engine is designed as a heat pump (i.e. the service rendered to the user consists of supplying heating power to the hot source) or as a refrigerator (i.e. the service rendered resides in the cooling power), a performance indicator will be defined respectively called the coefficient of amplification (COA, also sometimes called Energy Efficiency Rating, EER) and coefficient of performance (COP), and sometimes also called the cooling effect coefficient. The word "efficiency" will be avoided because both the COP and the COA can be greater than 1. They are defined as follows:

$$COA = \frac{-\dot{Q}_{cond}}{\dot{W}} \qquad [2.5]$$

$$COP = \frac{\dot{Q}_{evap}}{\dot{W}} \qquad [2.6]$$

These two quantities are by definition positive. In the absence of heat gains and rejections (adiabatic system, "ad"), note that $COA_{ad} = 1 + COP_{ad}$, so that under these conditions, $COA_{ad} > 1$.

In the presence of heat rejection and if the heat gains are negligible (typical configuration for heat pumps, which are generally at temperatures higher than the ambient temperature), then $\dot{Q}_p < 0$, and the COA can be rewritten as:

$$COA = \frac{\dot{W} + \dot{Q}_{evap} + \dot{Q}_p}{\dot{W}} = 1 + COP + \frac{\dot{Q}_p}{\dot{W}} < COA_{ad} \qquad [2.7]$$

It is more difficult to draw conclusions as to the impact of heat rejection or gain on refrigerators. Indeed, refrigerators are generally subject to both heat gain and rejection. It is then on the thermo-hydraulic coupling between the various components (some of which are rather the site of heat gains and the others heat rejection) that the system efficiency depends. If the heat rejection is relatively low compared to the heat gains (which implies that $\dot{Q}_p > 0$), whereas $COP < COP_{ad}$.

2.2.1.2. *Double-function heat pumps*

Figures 2.1 and 2.2 schematically represent reverse cycle engines independently of the service they render: heat pump or refrigerator. In some cases, for the user, a reverse cycle engine provides a double service: simultaneous cold and heat generation. This is commonly used in certain heat and cold consuming industries (e.g. in the food industry). Some urban areas have also chosen to combine two sports facilities: swimming pool (high heat consumption in winter) and ice rink (high cold consumption). Finally, the use of double-function heat pumps is growing in the building sector: the installation can provide some cooling while providing the heat required by domestic hot water.

Under these conditions, the energy performance indicator (EnPI) of a double-function heat pump can be defined as:

$$\text{EnPI} = \frac{\dot{Q}_{evap} - \dot{Q}_{cond}}{\dot{W}} = COP + COA = \frac{2\dot{Q}_{evap} + \dot{W} + \dot{Q}_p}{\dot{W}} = 1 + 2COP + \frac{\dot{Q}_p}{\dot{W}} \quad [2.8]$$

2.2.1.3. *Three-heat-source heat pumps and refrigerators*

To operate, three-heat-source heat pumps and refrigerators essentially absorb thermal power ($\dot{Q}_m$) from a heat source at the temperature T_m, but incidentally a certain mechanical power $\dot{W}$ a priori modest compared to the motive power of heat. The presence of heat rejection or gain cannot be excluded. Thus, the first law of thermodynamics applied to a three-heat-source reverse cycle engine such as that of Figure 2.3 operating in steady state is written as:

$$0 = \dot{W} + \dot{Q}_m + \dot{Q}_{abs} + \dot{Q}_{cond} + \dot{Q}_{evap} + \dot{Q}_p \quad [2.9]$$

By analogy with the definitions adopted for two-heat-source engines, and if the mechanical power is effectively negligible compared to the motive power of heat, then:

$$COA = \frac{-\dot{Q}_{cond} - \dot{Q}_{abs}}{\dot{Q}_m} \quad [2.10]$$

$$COP = \frac{\dot{Q}_{evap}}{\dot{Q}_m} \quad [2.11]$$

$$\text{EnPI} = \frac{\dot{Q}_{evap} - \dot{Q}_{cond} - \dot{Q}_{abs}}{\dot{Q}_m} \quad [2.12]$$

However, for some hybrid systems, for example, so-called compression–absorption systems, the mechanical power is not negligible and the following is adopted:

$$COA = \frac{-\dot{Q}_{cond} - \dot{Q}_{abs}}{\dot{W} + \dot{Q}_m}, \; COP = \frac{\dot{Q}_{evap}}{\dot{W} + \dot{Q}_m} \text{ and EnPI} = \frac{\dot{Q}_{evap} - \dot{Q}_{cond} - \dot{Q}_{abs}}{\dot{W} + \dot{Q}_m} \quad [2.13]$$

2.2.2. A system-wide analysis: power or energy?

Reasoning in powers at the macroscopic scale and for a system operating under steady state makes it easy for the design engineer to compare different options and to prioritize their merits. However, in practice, reverse cycle engines rarely operate in steady state. At the forefront of deviations is the necessary control of the system (on–off control, or a more progressive technology) to adapt to variations in ambient conditions, source temperatures and user needs. Other technological constraints may intervene, such as required defrosting periods.

Next, consider an integrated energy analysis over a period representative of the operation of the system Δt: on–off cycle if the system undergoes such cycles, day, heating season for a heat pump, cooling season for air conditioning, year for an industrial refrigeration plant, etc. Integrated over this duration, for a two-heat-source system, equation [2.3] is written as:

$$\Delta E = \int_0^{\Delta t} (\dot{W} + \dot{Q}_{cond} + \dot{Q}_{evap} + \dot{Q}_p) \mathrm{d}t = W + Q_{cond} + Q_{evap} + Q_p \quad [2.14]$$

On an on–off cycle, if the entire system is in the same state at the start and end of the cycle, ΔE is zero. Otherwise, over long durations, the total energy variation is generally negligible compared to the other energy terms of equation [2.14]. Provided that one of these conditions is met, it can therefore be accepted that:

$$W + Q_{cond} + Q_{evap} + Q_p = 0 \quad [2.15]$$

We can thus define a COP, a COA or an average EnPI:

$$COP_m = \frac{Q_{evap}}{W},\ COA_m = \frac{-Q_{cond}}{W} \text{ and } EnPI_m = \frac{Q_{evap} - Q_{cond}}{W} \quad [2.16]$$

It will be noted that these average performance indicators are calculated from the energies involved. They are not obtained by calculating their average integral value:

$$COP_m \neq \frac{1}{\Delta t} \int_0^{\Delta t} \frac{\dot{Q}_{evap}}{\dot{W}} dt,$$

$$COA_m \neq \frac{1}{\Delta t} \int_0^{\Delta t} \frac{-\dot{Q}_{cond}}{\dot{W}} dt \quad [2.17]$$

$$EnPI_m \neq \frac{1}{\Delta t} \int_0^{\Delta t} \frac{\dot{Q}_{evap} - \dot{Q}_{cond}}{\dot{W}} dt$$

2.2.3. *Component-scale energy analysis*

Reasoning by means of methodological illustration at the component scale, the energy analysis of a single-stage mechanical vapor compression engine (Chapter 1, section 1.1.2, Figure 1.3) will be discussed. This methodology can of course be adapted to the other systems described in the first chapter, since these different systems consist of the association, admittedly more complex, of the same thermofluidic or mechanical components (compressors, lines, heat exchangers, etc.).

2.2.3.1. *Thermodynamic cycle*

Different levels of complexity can be considered to describe and study the thermodynamic cycle of a single-stage vapor compression reverse cycle engine.

Level 1: at a minimum, the presence of the four main components of the system (compressor, expansion device, evaporator and condenser) implies thinking about the four components in all four states of the refrigerant at the four points of the cycle. It is typical to consider that the expansion device is adiabatic (which implies that the expansion of the fluid within it is isenthalpic), and reasoning on four points requires admitting that the lines are adiabatic. Finally, for an initial approach, the compressor is sometimes assumed to be adiabatic. These assumptions usually correspond to the case where $\dot{Q}_p = 0$.

Level 2: still reasoning on four points, taking into account the non-adiabatic nature of the compressor (which is generally realistic given the temperature level reached by the compressor) modifies the energy and entropy analyses. In this case, $\dot{Q}_p = \dot{Q}_{comp} < 0$ (heat rejection from the compressor to its surroundings).

Level 3: an additional level of complexity consists of adding in the analysis heat rejection from the discharge line (between the compressor and condenser) and heat gains to the suction line (between the evaporator and the compressor). The cycle can be described using six distinct points.

Level 4: heat rejection from the line located between the condenser and the expansion valve is considered, especially since a liquid reservoir is sometimes located at this point on certain systems, to allow the system to operate over a wide range of operating conditions.

Additional levels of complexity could be considered, to better take into account the operation of such and such a component or such and such a phenomenon, but such additions then generally only slightly affect the results of the energy, entropy and exergy analyses, even if such a level of complexity is not justified. For example, in practice, the expansion valve is located very close to the evaporator, and it would be difficult to assess heat exchanges in this section of line.

Finally, the analysis of the dynamic behavior of reverse cycle engines is a particularly complex task, for which methodologies still mostly remain to be invented. It is a question of establishing the relations between the instantaneous thermodynamic performances of the engine and the dynamic behavior of the components, which itself can generally only be well apprehended by the analysis of the dynamic behavior of the fluid and the metallic elements, from the viewpoint of heat flow (thermal inertia, thermal transients, etc.) and fluid mechanics (distribution of the fluid charge, pressure waves, etc.).

2.2.3.1.1. Level 1 energy analysis

The cycle corresponding to level 1 of complexity (as well as to level 2) is represented in a Mollier diagram ($\ln P$; h) in Figure 2.4, and point numbers are shown in the associated system diagram. This figure shows that the lines and the heat exchangers (condenser, evaporator) are the site of pressure drops. These pressure drops affect the results of the thermodynamic analyses, since the state functions (h, s) at the characteristic points of the cycle depend on the pressure at these points. In practice (section 2.2.3.2), generally because of insufficient instrumentation, simplifying assumptions must often be made, at the cost of significant errors in the numerical results of the analysis.

The application of the first law of thermodynamics (equation [2.2]) in steady state to each component of the cycle makes it possible to write, by denoting $\dot{m}$ the specific flow rate of constant and homogeneous refrigerant throughout the circuit:

$$\text{evaporator: } \dot{Q}_{evap} = \dot{m}(h_1 - h_4) \quad [2.18]$$

$$\text{condenser: } \dot{Q}_{cond} = \dot{m}(h_3 - h_2) \quad [2.19]$$

$$\text{expansion device: } h_3 = h_4 \quad [2.20]$$

$$\text{compressor: } \dot{W} = \dot{m}(h_2 - h_1) \quad [2.21]$$

Taking into account the definitions of the COP if the engine is considered as a refrigerator, and of the COA if it is a heat pump, then:

$$COP = \frac{h_1 - h_4}{h_2 - h_1}, \; COA = \frac{h_2 - h_3}{h_2 - h_1} \quad [2.22]$$

The following relationship between COP and COA is found if the system is adiabatic:

$$COA_{ad} = 1 + COP_{ad} \quad [2.23]$$

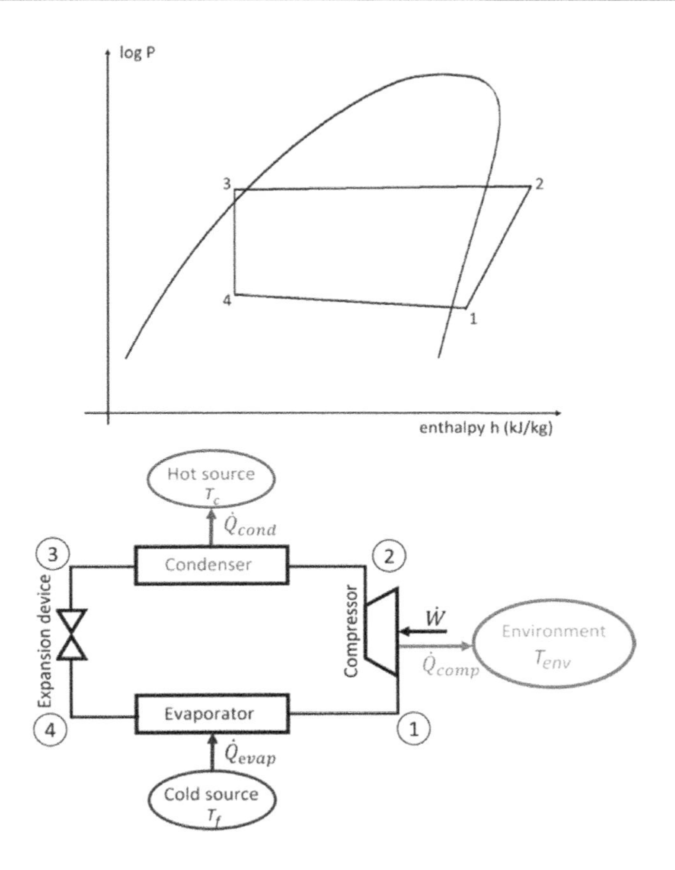

Figure 2.4. *Thermodynamic cycle of complexity of level 1 or level 2*

For a heat pump:

$$EnPI_{ad} = COP_{ad} + COA_{ad} = 1 + 2COP_{ad} \quad [2.24]$$

2.2.3.1.2. Level 2 energy analysis

In practice, during compression, the temperature of the compressed vapor increases, even for reversible compression, and a fortiori in the presence of irreversibilities. It follows that the temperature of the compressor also increases, so that this component is never adiabatic. Moreover, sometimes, it is even necessary for technological reasons (quality of the lubricating oil, for example) to limit the temperature of the compressor. A compressor cooling system is then implemented, for example, by means of fins to dissipate the heat to the ambient air by using an auxiliary fan, or even by means of water circulation. This is particularly the case when the refrigerant is ammonia.

Under these conditions, the equations expressing the first law of thermodynamics applied to the expansion device, the evaporator and the condenser (equations [2.18], [2.19] and [2.20]) remain unchanged, but for the compressor, the following is found:

$$\dot{W} = \dot{m}(h_2 - h_1) - \dot{Q}_{comp} \quad [2.25]$$

Since $\dot{Q}_{comp} < 0$, it can be deduced from this that for the same suction and discharge conditions, and for the same refrigerant flow rate, the heat rejected at the compressor increases the consumption of mechanical power, which also leads to a reduction in the COP, the COA and the EnPI, which are expressed as:

$$COP = \frac{\dot{m}(h_1 - h_4)}{\dot{m}(h_2 - h_1) - \dot{Q}_{comp}}$$

$$COA = \frac{\dot{m}(h_2 - h_3)}{\dot{m}(h_2 - h_1) - \dot{Q}_{comp}} \quad [2.26]$$

$$EnPI = \frac{\dot{m}(h_1 + h_2 - 2h_3)}{\dot{m}(h_2 - h_1) - \dot{Q}_{comp}}$$

The relationship between the COP and the COA is no longer immediate ($COA \neq COP + 1$) as a result of equation [2.7], the following is obtained:

$$COA = 1 + COP + \frac{\dot{Q}_{comp}}{\dot{W}} \quad [2.27]$$

2.2.3.1.3. Level 3 energy analysis

Figure 2.5 presents the thermodynamic cycle of complexity of level 3. The presence of heat rejection in the discharge line ($\dot{Q}_{ref} = \dot{m}(h_6 - h_2)$), heat input in the suction line ($\dot{Q}_{asp} = \dot{m}(h_1 - h_5)$) and heat rejection at the compressor ($\dot{Q}_{comp}$) requires entering points 5 (evaporator outlet) and 6 (condenser inlet). It is assumed in the following that all heat rejected is directed to the environment at a uniform temperature, which may not be verified for certain system architectures (e.g. air from a technical room and outside air at different temperatures).

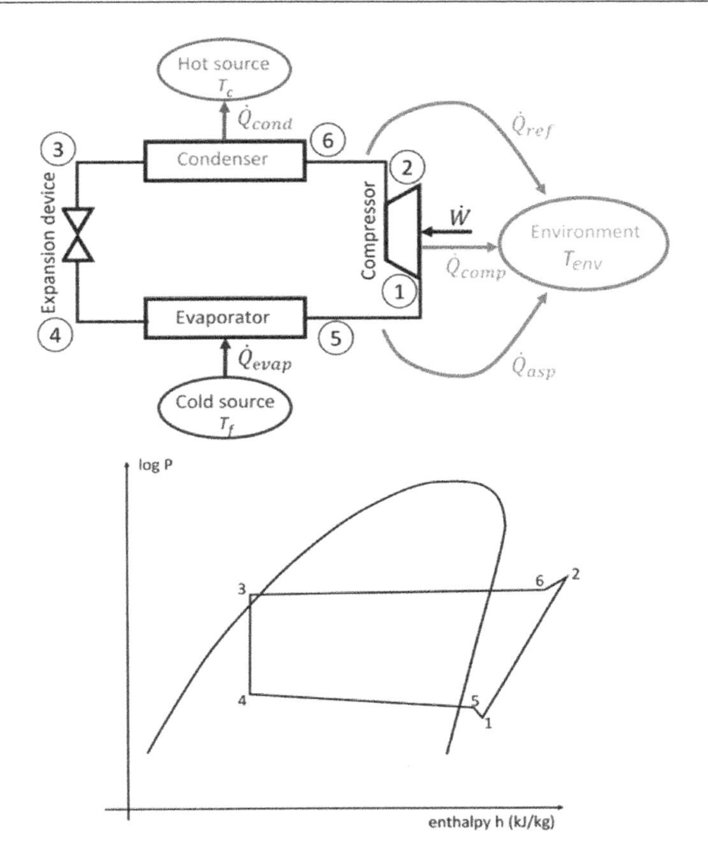

Figure 2.5. *Thermodynamic cycle of level 3 complexity*

The following powers are calculated:

$$\dot{Q}_{evap} = \dot{m}(h_5 - h_4)$$

$$\dot{Q}_{cond} = \dot{m}(h_3 - h_6) \qquad [2.28]$$

$$\dot{W} = \dot{m}(h_2 - h_1) - \dot{Q}_{comp}$$

and so:

$$COP = \frac{\dot{m}(h_5 - h_4)}{\dot{m}(h_2 - h_1) - \dot{Q}_{comp}}$$

$$COA = \frac{\dot{m}(h_6 - h_3)}{\dot{m}(h_2 - h_1) - \dot{Q}_{comp}} \quad [2.29]$$

$$EnPI = \frac{\dot{m}(h_5 + h_6 - 2h_3)}{\dot{m}(h_2 - h_1) - \dot{Q}_{comp}}$$

If points 3 and 4 are identical to those adopted for the level 2 analysis, and if points 5 and 6 are respectively identical to points 1 and 2 of the level 2 analysis (i.e. the inlets and outlets of the heat exchangers are unchanged), then it appears that the power consumed by the compressor in the presence of gains and losses in the lines is greater than that observed for level 2.

2.2.3.1.4. Level 4 energy analysis

At level 4 of complexity, a pressure drop and heat rejection at the condenser outlet are included, which leads to the cycle indicated in Figure 2.6.

In addition to the powers calculated previously, the thermal power linked to the rejection from the liquid line at the condenser outlet can be calculated: $\dot{Q}_{lsc} = \dot{m}(h_3 - h_7)$.

2.2.3.2. *Synthesis: what level of complexity for what detail of analysis?*

In this example, a refrigerator is intended to produce cold in a cold room at the temperature $T_f = 0°C$ with heat rejection to a source at $T_c = T_{amb} = 30°C$. The cooling capacity was rated at $\dot{Q}_f = 30$ kW by means of a heat balance on the cold source. The refrigerant is R1234ze(E), whose thermodynamic properties are available in the literature. Each level of complexity considered in section 2.2.3.1 corresponds to a number of measurements (pressure, temperature, power) carried out on the engine in operation. The conclusions that can be drawn from these measurements in terms of the energy behavior of the engine will be discussed.

Level 1

This level of complexity corresponds to the current technical practice: there are pressure measurements at the suction and discharge of the compressor, assumed to be respectively constant in the low-pressure parts (points 1 and 4), on the one hand, and high-pressure parts, on the other hand (points 2 and 3). This means that any effect induced by pressure drops in the circuit is disregarded, unfortunately without the possibility of evaluating the errors induced by this assumption. The temperature is also measured at the compressor suction, at the condenser outlet and, for heat pumps (but rarely for refrigerators), at the compressor discharge. The measured temperature and pressure values are shown in Table 2.1 in the gray boxes. The other values indicated in this table are deduced from the measurements. These are the saturation temperatures corresponding to these pressures as well as the enthalpies deduced from the coordinates (*T, P*). The enthalpy at the evaporator inlet is assumed to be equal to that characteristic of the condenser outlet, under the assumption of isenthalpic expansion.

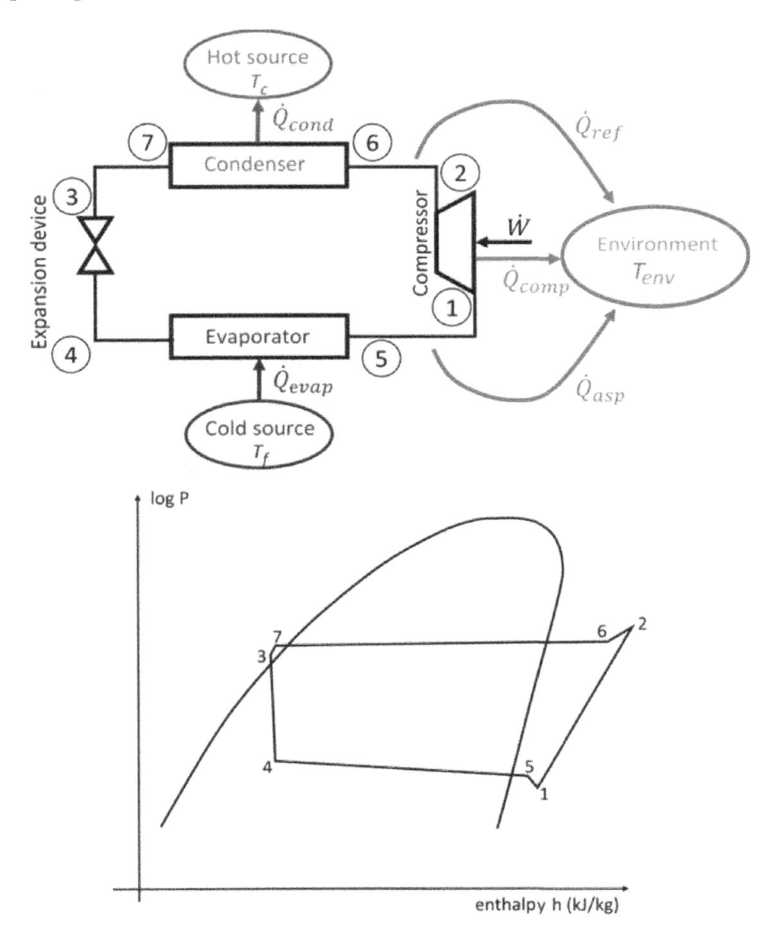

Figure 2.6. *Thermodynamic cycle of the complexity of level 4*

	Pressure P (bar)	Saturation temperature $T_{sat}(P)$ (°C)	Temperature T (°C)	Enthalpy h (kJ/kg)
(1) Suction	1.201	-15	-5	382.2
(2) Discharge	7.665	40	55	425.4
(3) Condenser outlet	7.665	40	35	247.8
(4) Evaporator inlet	1.201	-15	0	247.8

Table 2.1. *Properties of the fluid at the characteristic points of the cycle (levels 1 and 2 of complexity)*

The fluid flow is calculated as follows: $\dot{m} = \frac{\dot{Q}_f}{h_1 - h_4} = \frac{\dot{Q}_f}{h_1 - h_3} = 0.223 \text{ kg/s}$

The mechanical power to the compressor can also be deduced: $\dot{W} = \dot{m}(h_2 - h_1) = 9.64 \text{ kW}$

Finally, the COP can be calculated as: $COP = \frac{\dot{Q}_f}{\dot{W}} = 3.11$

Level 2

In addition to the temperature and/or pressure measurements at points 1, 2 and 3, this approach requires having a measurement of the power consumed by the compressor. The fluid flow evaluated is unchanged (since the characteristic points of the cycle are unchanged (Table 2.1), especially those of the evaporator) but the consumption appears to be higher than that calculated previously: experimentally, $\dot{W} = 11.0 \text{ kW}$. Thus, the heat rejected by the compressor represents more than 12% of the power consumed since $\dot{Q}_{comp} = \dot{m}(h_2 - h_1) - \dot{W} = -1.36 \text{ kW}$.

It follows that the COP is reduced to $COP = \frac{\dot{Q}_f}{\dot{W}} = 2.73$.

Level 3

Temperature and pressure measurements are added at points 5 and 6. The characteristic points of the cycle are given in Table 2.2.

These measurements reveal the pressure drops in the suction line and the temperature increases along it (pressure drops that were not accessible for analysis at levels of complexity 1 and 2), as well as heat rejected from the discharge line with a negligible pressure drop.

	Pressure P (bar)	Temperature T (°C)	Enthalpy h (kJ/kg)
(1) Suction	1.201	-5	382.2
(2) Discharge	7.665	55	425.4
(3) Condenser outlet	7.665	35	247.8
(5) Evaporator outlet	1.474	-5	381.5
(6) Condenser inlet	7.665	53	423.3

Table 2.2. *Properties of the fluid at the characteristic points of the cycle designed at level 3 of complexity*

In addition, it is noted that the measurement of the mechanical power is not significantly modified ($\dot{W} = 11.0$ kW).

It therefore appears that the fluid flow $\dot{m} = \frac{\dot{Q}_f}{h_5 - h_4} = \frac{\dot{Q}_f}{h_5 - h_3} = 0.224\ \text{kg/s}$ is practically identical to that determined at level 1, so that the heat rejection remains close to that determined at level 2: $\dot{Q}_{comp} = \dot{m}(h_2 - h_1) - \dot{W} = -1.31$ kW.

The COP remains unchanged: $COP = \frac{\dot{Q}_f}{\dot{W}} = 2.73$.

It is also possible to calculate the thermal power rejected at the hot source:

$$\dot{Q}_{cond} = \dot{m}(h_3 - h_6) = -39.38\ \text{kW}.$$

If this engine were used as a heat pump, it would show a COA equal to 3.58.

This makes it possible to evaluate the heat gained and rejected at the lines, $\dot{Q}_{asp} = \dot{m}(h_1 - h_5) = 0.16$ kW and $\dot{Q}_{ref} = \dot{m}(h_6 - h_2) = -0.47$ kW.

These values are negligible compared to the other thermal powers of the cycle. The analysis at level 2 seems sufficient to correctly evaluate the energy behavior of the system.

Level 4

Finally, always with $\dot{W} = 11.0$ kW, note the points given in Table 2.3.

The only difference compared to level 3 lies in the possibility of evaluating the heat rejection at the liquid line at the condenser outlet $\dot{Q}_{lsc} = \dot{m}(h_3 - h_7) = -0.34$ kW.

	Pressure P (bar)	Temperature T (°C)	Enthalpy h (kJ/kg)
(1) Suction	1.201	-5	382.2
(2) Discharge	7.665	55	425.4
(3) Expansion device inlet	7.458	35	247.8
(5) Evaporator outlet	1.474	-5	381.5
(6) Condenser inlet	7.665	53	423.3
(7) Condenser outlet	7.665	36	249.3

Table 2.3. *Properties of the fluid at the characteristic points of the cycle designed at level 4 of complexity*

This confirms that from the energy point of view, the analysis at level 2 seems quite sufficient.

2.3. Entropy analysis

Entropy analysis aims to identify and quantify the imperfections of reverse cycle engines, assessing the irreversibilities and ultimately the deviation from the theoretical upper limit in terms of performance that ideal reversible operation would represent.

There are multiple sources of irreversibility in the operation of a reverse cycle engine. These may be external irreversibilities, linked to heat transfer between the engine and the heat sources with which it exchanges heat (hot source, cold source, external environment, etc.). It can also be internal phenomena: mechanical irreversibilities (friction, shocks, pressure drops during flows, phenomena specific to two-phase flows with phase change, etc.), chemical (operation of mixing or separation of fluids in certain applications such as sorption engines as well as vapor compression engines for which the refrigerant is a multicomponent fluid), etc. Certain irreversibilities sometimes deserve to be evaluated by means of local entropy analyses, because such an analysis can turn out to be rich in lessons, but this is often time-consuming and not always justified. Thus, in this part, some entropy analysis approaches that are often useful will be discussed, without aiming to be exhaustive. Beforehand, some reminders of thermodynamics relating to the second law are proposed.

2.3.1. *Second law of thermodynamics: an entropic power balance*

The second law of thermodynamics applied to a closed system can be written as an entropic power balance:

$$\frac{dS}{dt} = \sum \frac{\dot{Q}_i}{T_i} + P(s) \quad \text{with} \quad P(s) \geq 0 \qquad [2.30]$$

S represents here the entropy of the system, $\dot{Q}_i$ refers to all the thermal powers exchanged by the system with an external source at temperature T_i, and $P(s)$ represents the entropy generation (rate of entropy generation) in the system, which is a measure of the irreversibilities generated by the system during its operation.

For a flow of matter through an open system, the second law of thermodynamics reads:

$$\frac{dS}{dt} = \sum \frac{\dot{Q}_i}{T_i} + \sum(\dot{m}_e s_e - \dot{m}_s s_s) + P(s) \qquad [2.31]$$

where $\dot{m}_e$ denotes entering specific flow rates, $\dot{m}_s$ denotes exiting specific flow rates, s_e denotes the specific entropy of the entering matter flows and s_s denotes the specific entropy of exiting matter flows. The same sign conventions as for equations [2.1] and [2.2] are adopted.

2.3.2. *Reversible upper limit: Carnot engines*

2.3.2.1. *Two-heat-source reverse cycle engine*

2.3.2.1.1. External hot and cold sources

The second law of thermodynamics is applied in order to carry out an entropic power balance on the system of Figure 2.2. If only the two hot and cold sources exchange heat with the system (i.e. $\dot{Q}_p = 0$) and under steady state, this balance is written as:

$$0 = \frac{\dot{Q}_{cond}}{T_c} + \frac{\dot{Q}_{evap}}{T_f} + P(s) \qquad [2.32]$$

Combining this relationship with the first law (equation [2.4], with $\dot{Q}_p = 0$), we can substitute $\dot{Q}_{cond} = -\dot{W} - \dot{Q}_{evap}$ and write the following:

$$\dot{Q}_{evap}\left(\frac{1}{T_f} - \frac{1}{T_c}\right) - \frac{\dot{W}}{T_c} + P(s) = 0 \qquad [2.33]$$

Taking into account its definition, the COP (equation [2.6]) becomes:

$$COP = \frac{\dot{Q}_{evap}}{\dot{W}} = \frac{1/T_C}{\frac{1}{T_f} - \frac{1}{T_C}} - \frac{P(s)}{\left(\frac{1}{T_f} - \frac{1}{T_C}\right)\dot{W}} = \frac{T_f}{T_C - T_f} - \frac{P(s)}{\left(\frac{1}{T_f} - \frac{1}{T_C}\right)\dot{W}} \qquad [2.34]$$

If the system is reversible, $P(s) = 0$ and the upper COP limit is deduced, which is called the Carnot COP to emphasize that this is the COP that the engine would have if it were reversible, in the absence of heat rejection or gain, and if it only exchanged heat with the two hot and cold sources:

$$COP_{Carnot} = \left.\frac{\dot{Q}_{evap}}{\dot{W}}\right]_{P(s)=0} = \frac{T_f}{T_C - T_f} \qquad [2.35]$$

The two previous equations clearly show that it is the irreversibilities that degrade the COP from its upper limit to its real value.

In a similar way, to define the other performance indicators (COA and EnPI), the corresponding Carnot indicators are defined ("Carnot COA", "Carnot EnPI") and written as follows:

$$COA_{Carnot} = \frac{T_C}{T_C - T_f} \qquad [2.36]$$

$$EnPI_{Carnot} = \frac{T_C + T_f}{T_C - T_f} \qquad [2.37]$$

The most immediate way to quantify the deviation from ideality of a two-heat-source reverse cycle engine, including if it is also subject to heat gain and rejection, consists of reporting the appropriate performance indicator (COP, COA or EnPI) to its maximum Carnot limit. This ratio is generally called thermodynamic efficiency, or for refrigerators, sometimes called refrigeration efficiency:

For a refrigerator, $\eta = \frac{COP}{COP_{Carnot}}$

For a heat pump, $\eta = \frac{COA}{COA_{Carnot}}$ [2.38]

For a double-function heat pump, $\eta = \frac{EnPI}{EnPI_{Carnot}}$

This efficiency value is by definition between 0 and 1 and it is therefore immediately interpretable, which is a clear advantage over its primary indicator. As a matter of fact, the upper limit of the primary indicator is its Carnot limit, which itself depends non-intuitively on the source temperatures. Thus, depending on the context (depending on the source temperature values), the same COP (or COA or EnPI) value may turn out to be close to the upper limit, or much further away, which will be perfectly reflected by an efficiency respectively close to 1 or 0.

2.3.2.1.2. Carnot cycle based on equivalent internal sources

Defining the Carnot indicators from the temperatures of the sources (external to the reverse cycle engine) includes in the analysis the external irreversibilities associated with the heat transfer with the external sources. However, these can be high, which can affect our understanding of how internal irreversibilities are distributed. To eliminate external irreversibilities from the analysis, we can then define equivalent internal sources, which will be used to determine the Carnot indicators of the cycle (and no longer of the engine associated with the external sources). It is then advisable to choose these temperatures of equivalent internal sources judiciously, under penalty of introducing certain biases underlined in this section and the following one.

A simple initial approach considers that the equivalent internal hot source is at the condensation temperature and the cold source temperature at the evaporation temperature. The justification, of course imperfect from a thermodynamic point of view, lies in the very high heat transfer coefficients associated with phase change processes. If the temperature is not constant during the phase change (significant pressure drop of a pure refrigerant, or temperature glide induced by the zeotropic nature of a multicomponent refrigerant), but the difference temperature between the inlet and outlet of the component remains modest (a few Kelvin, for example), it is often accepted that the arithmetic mean between the inlet and output temperatures is sufficient to define the equivalent internal temperature. Under these conditions, the Carnot indicators are expressed as:

$$
\begin{aligned}
COP_{Carnot} &= \frac{T_{evap}}{T_{cond}-T_{evap}} \\
COA_{Carnot} &= \frac{T_{cond}}{T_{cond}-T_{evap}} \qquad [2.39] \\
EnPI_{Carnot} &= \frac{T_{cond}+T_{evap}}{T_{cond}-T_{evap}}
\end{aligned}
$$

However, this approach is not very satisfactory because it does not take into account the internal irreversibilities that occur in the condenser when the refrigerant is unsuperheated, and to a lesser extent when the liquid is subcooled at the outlet of the condenser. Likewise, the necessary superheating of the vapor that occurs near the outlet of the evaporator also leads to irreversibilities that should be included in the search for an equivalent cycle. In other words, this approach eliminates all external irreversibilities, as well as those that are probably modest (phase change) as well as those that may be more significant (heat exchanges while the refrigerant is in a single-phase state, characterized by more resistance to heat transfer).

To choose these internal equivalent source temperatures, we must thus provide a definition that guarantees a greater thermodynamic coherence, which is possible with the concept of entropic temperature. For any transformation between state 1 and state 2, the entropic temperature is defined as:

$$\tilde{T}_{12} = \frac{\Delta h_{12}}{\Delta s_{12}} \quad [2.40]$$

Therefore, $\tilde{T}_{cond} = \frac{\Delta h_{inlet;outlet\ condenser}}{\Delta s_{inlet;outlet\ condenser}}$ and $\tilde{T}_{evap} = \frac{\Delta h_{inlet;outlet\ evaporator}}{\Delta s_{inlet;outlet\ evaporator}}$, and the relations given by equation [2.39] can be used by replacing the evaporation and condensation temperatures with the corresponding entropic temperatures.

Figure 2.7 schematically represents, on a temperature scale, the temperature profiles in the exchangers and the relative position of the source temperatures (external or internal). This representation helps to interpret the Carnot COP values since the Carnot COP is inversely proportional to the difference between the temperatures of the sources.

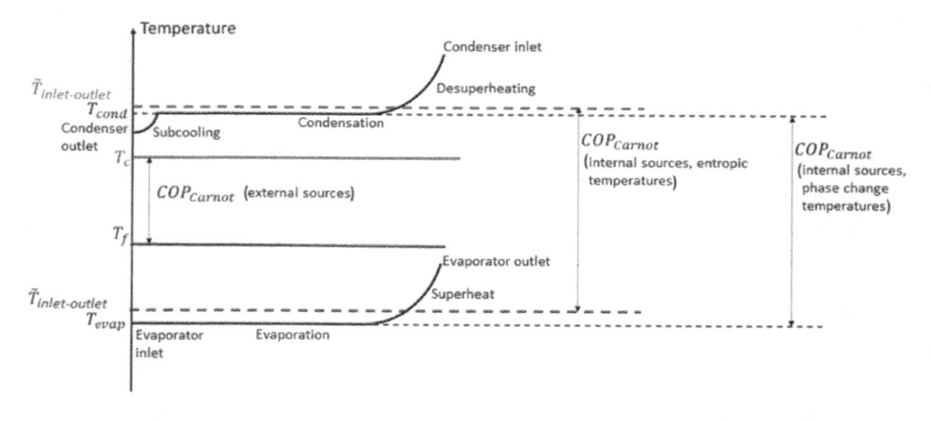

Figure 2.7. *Schematic representation of the temperatures of equivalent external or internal sources and temperature profiles in the exchangers*

2.3.2.1.3. Summary: external sources or internal sources?

The refrigerator discussed in section 2.2.3.1 is considered here with respect to the levels of complexity 1, 2 and 3. The characteristics of the cycle must be completed by the entropies (Table 2.4 for level 2, Table 2.5 for level 3), which makes it possible to calculate the characteristic entropic temperatures of the evaporator and the condenser. From these data, the following are compared in Table 2.6: the Carnot COP and cooling values estimated according to the level of complexity (chosen by the analyst), and according to their choice of temperatures of equivalent external or internal sources. These data are graphically translated in Figure 2.8.

This synthesis firstly reveals the importance of external irreversibilities, linked to heat transfers between the engine and external sources. The Carnot COP calculated from the temperatures of internal sources appears here to be almost half of the Carnot COP calculated from the external sources.

In addition, the use of phase change temperatures instead of entropic temperatures seems quite reasonable – it should nevertheless be emphasized that this remains true for moderate vapor superheats at the compressor discharge port: the greater the discharge superheat, the lesser the entropic temperature at the condenser is close to the condensation temperature.

	Pressure P (bar)	Phase change temperature $Tsat(P)$ (°C) / (K)	Temperature T (°C) / (K)	Enthalpy h (kJ/kg)	Entropy s (kJ/kg.K)	Entropic temperature (K)
Discharge (2)	7.665	40 / 313	55 / 328	425.4	1.729	313.2
Condenser outlet (3)	7.665	40 / 313	35 / 308	247.8	1.162	
Evaporator inlet (4)	1.201	-15 / 258	-5 / 268	247.8	1.188	258.4
Suction (1)	1.201	-15 / 258	-5 / 268	382.2	1.708	

Table 2.4. *Properties of the fluid at the characteristic points of the cycle (levels 1 and 2 of complexity)*

	Pressure P (bar)	Phase change temperature $Tsat(P)$ (°C) / (K)	Temperature T (°C) / (K)	Enthalpy h (kJ/kg)	Entropy s (kJ/kg.K)	Entropic temperature (K)
Suction (1)	1.201	-15 / 258	-5 / 268	382.2	1.708	N/A
Discharge (2)	7.665	40 / 313	55 / 328	425.4	1.728	N/A
Condenser inlet (6)	7.665	40 / 313	53 / 326	423.3	1.722	313.4
Condenser outlet (3)	7.665	40 / 313	35 / 308	247.8	1.162	
Evaporator inlet (4)	1.474	-10 / 263	-10 / 263	247.8	1.183	263.2
Evaporator outlet (5)	1.474	-10 / 263	-5 / 268	381.5	1.691	

Table 2.5. *Properties of the fluid at the characteristic points of the cycle designed at level 3 of complexity*

	Level 1 (COP = 3.11)		Level 2 (COP = 2.73)		Level 3 (COP = 2.73)	
	COP_{Carnot}	η	COP_{Carnot}	η	COP_{Carnot}	η
External source temperatures	9.10	0.341	9.10	0.300	9.10	0.300
Phase change temperatures	4.69	0.663	4.69	0.582	5.26	0.519
Entropic temperatures	4.72	0.660	4.72	0.579	5.24	0.521

Table 2.6. *Carnot COP and cooling efficiency according to the analyst choice in terms of the level of complexity and source temperatures*

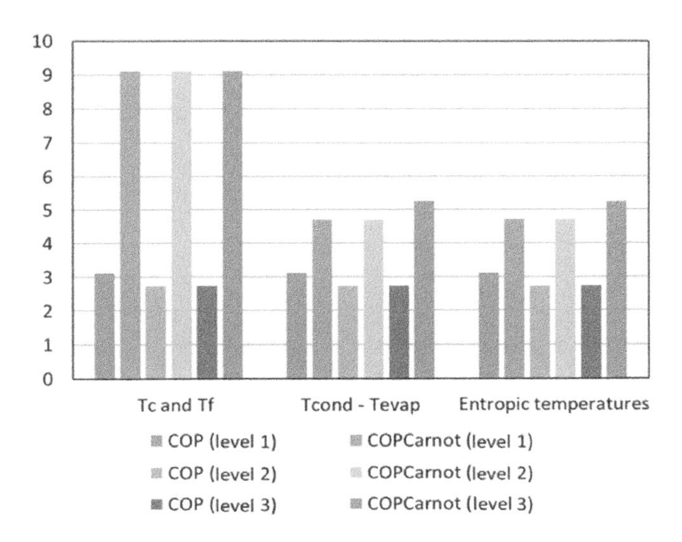

Figure 2.8. *Influence of analysis choices on Carnot COP and refrigeration efficiency*

Finally, a complexity of level 3, which did not appear essential for the energy analysis (see section 2.2.3.2), leads to significantly different cooling efficiency values from level 2 complexity if the internal irreversibilities are specifically studied while excluding external irreversibilities. For a detailed entropy analysis, the complexity of level 3 is therefore preferred.

2.3.2.1.4. Concept of entropy generation number

This concept was introduced by Professor Adrian Bejan in the early 1980s (Bejan, 1982). The entropy generation number is defined as the ratio between the entropy generation of the reverse cycle engine and the maximum entropy generation that would be obtained if all of the mechanical power were degraded completely irreversibly at room temperature; for example, by converting it (100% efficiency) to electricity, which is then dissipated by the Joule effect in an electrical resistor. By imagining this system, the first and the second laws of thermodynamics read:

$$\dot{Q} = -\dot{W} \qquad [2.41]$$

and

$$P(s)_{max} = -\frac{\dot{Q}}{T_{amb}} = \frac{\dot{W}}{T_{amb}} \qquad [2.42]$$

The entropy generation number is then defined as:

$$N_s = P(s)/P(s)_{max} \quad [2.43]$$

It is easily shown that if a refrigerator is not subjected to heat gains and if the ambient temperature is equal to the hot source temperature, the thermodynamic efficiency is equal to:

$$\eta = 1 - N_s \quad [2.44]$$

The same is true for a heat pump for which $\dot{Q}_p = 0$, and if $T_{amb} = T_f$.

2.3.2.2. *Three-heat-source reverse cycle engine*

As for the two-heat-source system engine, the second law of thermodynamics is applied in order to carry out an entropic power balance on the system of Figure 2.3. In steady state and in the absence of heat gain and rejection, this entropy balance is written as:

$$0 = \frac{\dot{Q}_{cond}+\dot{Q}_{abs}}{T_C} + \frac{\dot{Q}_{evap}}{T_f} + \frac{\dot{Q}_{m}}{T_m} + P(s) \quad [2.45]$$

2.3.2.2.1. Upper limit without mechanical power

From the theoretical point of view, the consumption of mechanical power is not absolutely necessary and this power is only a technological artifact, which helps the operation of the cycle without it being strictly necessary. We could also imagine theoretically that this power is itself produced using a heat engine (Carnot engine for the reversible upper limit), operating between the sources at the motive temperature and the hot source, thus eliminating this power from the energy balance. Under this assumption, the first and second laws are reduced to these forms:

$$\dot{Q}_{cond} + \dot{Q}_{abs} + \dot{Q}_{evap} + \dot{Q}_m = 0 \quad [2.46]$$

$$\frac{\dot{Q}_{cond}+\dot{Q}_{abs}}{T_C} + \frac{\dot{Q}_{evap}}{T_f} + \frac{\dot{Q}_{m}}{T_m} + P(s) = 0 \quad [2.47]$$

The Carnot COP, which corresponds to the reversible upper limit, can then be easily expressed as follows:

$$COP_{Carnot} = \left.\frac{\dot{Q}_{evap}}{\dot{Q}_{m}}\right]_{P(s)=0} = \frac{\frac{1}{T_C}-\frac{1}{T_m}}{\frac{1}{T_f}-\frac{1}{T_m}} = \frac{1-\frac{T_C}{T_m}}{\frac{T_C}{T_f}-1} \quad [2.48]$$

This approach, having neglected heat gain and rejection and excluded mechanical power from the reasoning, makes it possible to identify the relationship between the Carnot COP, COP and irreversibilities (quantified by the entropy generation). Indeed, in the case of a non-zero entropy generation, we can demonstrate that:

$$COP = \left.\frac{\dot{Q}_{evap}}{\dot{Q}_{\mathrm{m}}}\right]_{P(s)\neq 0} = COP_{Carnot} - \frac{T_C.P(s)}{\left(\frac{T_C}{T_f}-1\right)\dot{Q}_m} \quad [2.49]$$

The COP of the real system (non-reversible) corresponds to the Carnot COP from which a value proportional to the measure of the irreversibilities is deduced, i.e. the entropy generation.

In the absence of heat gain and rejection, the relationships between COP, COA and EnPI (like those between their upper reversible Carnot limits) remain valid:

$$COA = COP + 1, EnPI = COA + COP, COA_{Carnot} = COP_{Carnot} + 1 \quad [2.50]$$

2.3.2.2.2. Upper limit including mechanical power

However, mechanical power is not always a technological artifact. For example, certain so-called hybrid or combined systems, such as those based on absorption–compression cycles, actually require mechanical power, which is not absolutely negligible, and in any case which is intrinsic to the very operating principle of the cycle. The first and second laws of thermodynamics can then be rewritten as follows:

$$\dot{Q}_{cond} + \dot{Q}_{abs} + \dot{Q}_{evap} + \dot{Q}_m + \dot{W} = 0 \quad [2.51]$$

$$\frac{\dot{Q}_{cond}+\dot{Q}_{abs}}{T_C} + \frac{\dot{Q}_{evap}}{T_f} + \frac{\dot{Q}_{\mathrm{m}}}{T_m} + P(s) = 0 \quad [2.52]$$

Given that the mechanical power remains low compared to the thermal motive power, the COP can continue to be defined as $COP = \left.\frac{\dot{Q}_{evap}}{\dot{Q}_{\mathrm{m}}}\right]_{P(s)\neq 0}$ and hence:

$$COP = \frac{\frac{1}{T_C}-\frac{1}{T_m}}{\frac{1}{T_f}-\frac{1}{T_m}} - \frac{T_C.P(s)}{\left(\frac{T_C}{T_f}-1\right)\dot{Q}_m} + \frac{\dot{W}}{\left(\frac{T_C}{T_f}-1\right)\dot{Q}_m} \quad [2.53]$$

The reversible upper limit of the COP is therefore:

$$COP_{Carnot} = \frac{\frac{1}{T_C}-\frac{1}{T_m}}{\frac{1}{T_f}-\frac{1}{T_m}} + \frac{\dot{W}}{\left(\frac{T_C}{T_f}-1\right)\dot{Q}_m} \quad [2.54]$$

The comparison of this relation with that of equation [2.48] shows that at the cost of additional mechanical energy consumption, the system efficiency is still likely to increase, since its reversible upper limit is increased by the second term of the right-hand side.

If the mechanical power if disregarded to define the COP by adopting $COP = \left.\frac{\dot{Q}_{evap}}{\dot{Q}_{\mathrm{m}}+\dot{W}}\right]_{P(s)\neq 0}$, mathematical relationships are not noteworthy.

2.3.3. *Component-scale entropy analysis*

The entropy analysis presented previously makes it possible to obtain global indications as to the performance of the systems studied, in particular by comparing the COP (or COA, or EnPI) with its maximum theoretical limit obtained for a reversible system and exchanging heat only with its essential sources (hot, cold and possibly motive), excluding heat rejection and gain. If the deviation from reversibility is significant, the analyst then seeks to determine which component(s) and which phenomenon(a) contribute the most to overall irreversibilities: it is therefore generally the distribution of irreversibilities within the system that needs to be determined, whether or not the system is subject to heat gain and rejection.

This method of analysis will be introduced by continuing the study of the refrigerating system presented above, for its steady state, retaining the design at level 3 of complexity. If level 1 is a priori sufficient to describe the energy functioning of the system, level 3 differs significantly from level 2 as soon as an entropy analysis is done.

2.3.3.1. *Compressor, expansion device, circulation lines*

The second law of thermodynamics (see equation [2.31]) can be applied to each component and to each line where there are irreversibilities of thermal (heat gain or rejection from/to the outside) or mechanical (pressure drops) origin. Here (Figure 2.5):

$$\text{Compressor}: \frac{\dot{Q}_{comp}}{T_{amb}} + \dot{m}(s_1 - s_2) + P_{comp}(s) = 0 \quad [2.55]$$

Discharge line: 2→6 : $\frac{\dot{Q}_{ref}}{T_{amb}} + \dot{m}(s_2 - s_6) + P_{ref}(s) = 0$ [2.56]

Expansion device: $\dot{m}(s_3 - s_4) + P_{dét}(s) = 0$ [2.57]

Suction line 5→1 : $\frac{\dot{Q}_{asp}}{T_{amb}} + \dot{m}(s_5 - s_1) + P_{asp}(s) = 0$ [2.58]

2.3.3.2. *Heat exchangers: evaporator and condenser*

The duality introduced in section 2.3.2.1.2 as for the choice of the temperature of the sources takes on particular importance here. Indeed, it is possible to consider excluding external irreversibilities by choosing the entropic temperature as the equivalent source temperature "seen" by the exchanger. In this case:

Condenser: $\frac{\dot{Q}_{cond}}{\tilde{T}_{63}} + \dot{m}(s_6 - s_3) + P_{cond}(s) = 0$ [2.59]

Evaporator : $\frac{\dot{Q}_{évap}}{\tilde{T}_{45}} + \dot{m}(s_4 - s_5) + P_{évap}(s) = 0$ [2.60]

But since $\dot{Q}_{cond} = \dot{m}(h_3 - h_6)$, $\dot{Q}_{evap} = \dot{m}(h_5 - h_4)$ and given the definition of the entropic temperature (equation [2.40]), these equations lead to $P_{cond}(s) = 0$ and $P_{évap}(s) = 0$. This observation constitutes another illustration of the concept of equivalent internal source temperature, which makes it possible to exclude internal irreversibilities in the heat exchangers to show these irreversibilities entirely outside the system (no internal resistance to heat transfer; resistance heat transfer entirely on the side of the external source). Thus, this choice does not provide information on the contribution of exchangers to irreversibilities, which of course reduces the scope of an analysis at the component level.

For such an analysis, it therefore seems more judicious to reason on the temperature of external sources, and thus to include, in the entropy generation which will be evaluated for the exchangers, the irreversibilities linked to heat transfers to the hot source or from the cold source:

Condenser: $\frac{\dot{Q}_{cond}}{T_C} + \dot{m}(s_6 - s_3) + P_{cond}(s) = 0$ [2.61]

Evaporator: $\frac{\dot{Q}_{evap}}{T_f} + \dot{m}(s_4 - s_5) + P_{evap}(s) = 0$ [2.62]

It is common for heat transfer on the side of the external source to result in the heating of a heat transfer fluid (e.g. condenser using outside air as a heat carrier, the latter heating up in the condenser) or the cooling of a secondary fluid (e.g. heat extracted from the outside air (secondary fluid) for an air–air heat pump). Here, $\dot{m}_{fl}$ is the flow rate of the refrigerant or the heat carrier, s_{in} is its specific entropy at its inlet into the exchanger, and s_{out} is its specific entropy at the outlet; we can calculate the entropy generation in the exchanger:

$$P_{cond\ or\ evap}(s) = \dot{m}\Delta s_{63\ or\ 45} + \dot{m}_{fl}\Delta s_{in;out} \qquad [2.63]$$

If the secondary fluid or the heat carrier fluid can be considered either as an incompressible liquid or as an ideal gas whose pressure variation is negligible (which is usually the case for common industrial or commercial refrigeration applications, as well as for heat pumps for the heating of occupied premises), with c_p being the specific heat capacity of the fluid, therefore:

$$P_{cond\ or\ evap}(s) = \dot{m}\Delta s_{63\ or\ 45} + \dot{m}_{fl}c_p\ln\left(\frac{T_{out}}{T_{in}}\right) \qquad [2.64]$$

2.3.3.3. *Application and comments*

2.3.3.3.1. Distribution of irreversibilities

The relevant refrigerant properties are provided in Table 2.5. It can be recalled that (see section 2.2.3.2) $\dot{m} = 0.224\ \text{kg/s}$, $\dot{Q}_{comp} = -1.31\ \text{kW}$, $\dot{Q}_{cond} = -39.38\ \text{kW}$, $\dot{Q}_{asp} = 0.16\ \text{kW}$ and $\dot{Q}_{ref} = -0.47\ \text{kW}$.

So far, the ambient temperature has not been discussed. Ideally, the ambient temperature should be evaluated individually for each component and each circulation line, to appear in the numerator of the terms in $\dot{Q}/T_{amb}$. Indeed, the compressor can be in a technical room, the discharge line in the open air, the suction line partially in the cold room, etc. Each component thus experiences a different external environment. Otherwise, in refrigeration applications, it is common for this ambient temperature to be the hot source temperature if the ambient air is used as the heat carrier at the condenser. Similarly, for heat pump applications, the ambient air is often used as the secondary fluid at the evaporator, so that the cold source temperature and the ambient temperature are identical. These considerations are less justifiable if the secondary fluid or the heat carrier is water, or even sometimes a fluid contained in a fluidic loop independent of the ambient air (groundwater; water–glycol mixture loop; etc.). The analyst will therefore ensure the realistic nature of the temperatures chosen for the irreversibilities linked to the heat exchanges. It is assumed here that $T_{amb} = T_c = 30°\text{C} = 303\ \text{K}$ and $T_f = 0°\text{C} = 273\ \text{K}$.

The entropy generation associated with each component is calculated using the preceding equations. The values are grouped in Table 2.7, and presented as a pie chart in Figure 2.9.

	$P(s)$ (W/K)
Suction	3.296
Compressor	8.800
Discharge	0.194
Condenser	4.325
Expansion device	4.712
Evaporator	4.096
Total	**25.42**

Table 2.7. *Distribution of the entropy generation in the different components and circulation lines of the refrigeration system studied*

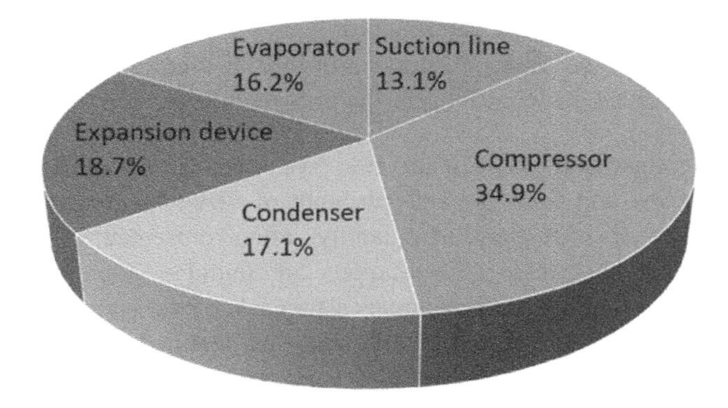

Figure 2.9. *Contribution of components and circulation lines to the total entropy generation of the studied system*

These values deserve several comments. Note first of all the predominance of the irreversibilities associated with the compressor, quite typical in this type of reverse cycle engine. This component is both the site of a significant heat release (more than 12% in this case; see section 2.3.2) and a mechanically very irreversible compression process (friction, shock).

The relatively strong contribution of the expansion valve is explained by the fact that, in addition to the intrinsically very irreversible nature of an expansion operation (pressure drops), the process of phase change during a flow in an organ, which restricts the cross-sectional area of the fluid, results in irreversibilities, which are greater for higher values of vapor quality. On the one hand, an expansion process is intrinsically highly irreversible, and in addition, in the present case, the fluid leaves the expansion device in the two-phase state with a relatively high vapor quality, of approximately 0.32. To reduce this source of irreversibility, we could seek to further subcool the refrigerant at the outlet of the condenser (e.g. by increasing the heat transfer area of the condenser), which would reduce the vapor quality at the outlet of the expansion device. If the other points of the cycle were unchanged and if the flow rate were kept constant, this would lead to an increase in the cooling capacity (increase in the enthalpy variation between the inlet and the outlet of the evaporator), in the heat rejection, as well as to an increase (albeit of a lower extent) in the entropy generation at the condenser and at the evaporator. Thus, overall, increasing subcooling only moderately reduces the entropy generation of the full system. Moreover, the more the subcooling increases, the less the total entropy generation of the system is significantly reduced, so that it becomes of little benefit to exceed a subcooling of a few Kelvin.

The discharge line is the site of heat rejection without a pressure drop, while the suction line is the site of more modest heat gains, but with a pressure drop of approximately 10% of the upstream pressure value. It therefore appears that it is rather the pressure drops that contribute to the irreversibilities.

Finally, the comparison between the first two levels of complexity (from the energy point of view and from the entropy point of view) leads us to discuss the concept of isentropic efficiency, a quantity often presented as an intrinsic characteristic of the compressor. Correlations are found in the literature which allegedly predict the isentropic efficiency as a function of the operating parameters of the compressor. This isentropic efficiency is defined, *for an adiabatic compressor*, as the ratio between the enthalpy variation of the refrigerant through the compressor if the compression was isentropic and the actual enthalpy variation:

$$\eta_{is} = \frac{h_{2s}-h_1}{h_2-h_1} \qquad [2.65]$$

Under these conditions (if the compressor can effectively be considered as adiabatic, in particular if the heat rejection or gain are negligible compared to the mechanical power), this efficiency is an indicator of the deviation from the reversibility of the compression. It is also easy to use since it suffices to evaluate the enthalpies (a priori according to the pressure and the temperature at the suction and discharge of the compressor) and to determine the isentropic output conditions to calculate it.

But in practice, the compressor of a refrigeration system is generally not adiabatic, and in this case, the isentropic efficiency loses its meaning by not making it possible to distinguish the share of work on the shaft actually transmitted to the fluid from that converted into heat. Taking into account equations [2.55] and [2.25] indicates that for a given mechanical power, if the heat rejection increases (in absolute value, i.e. if $|\dot{Q}_{comp}|$ increases), then the enthalpy at the compressor outlet decreases, which results in an increase in the isentropic efficiency as defined (then erroneously) by equation [2.65], to the point that during a cooled vapor compression (e.g. cooling of an ammonia compressor with a high water flow rate), the isentropic efficiency may even appear to be greater than 1! The isentropic efficiency is in no way intrinsic to the compression or to the compressor, but rather extrinsic: as defined carelessly by equation [2.65], it is above all a metric for the heat rejected from the compressor to its environment. If the heat rejection is included in the reflection (level 2 and more), consistency thus requires us not to include the isentropic efficiency in the analysis.

2.3.3.3.2. Entropy generation number

The concept of entropy generation numbers will be implemented for the studied system operating as a refrigerator, as well as assuming that it is used as a heat pump. In the latter case, the entropy generations are different from those presented in Table 2.7 in the sense that the ambient temperature towards which the heat rejection is directed is the cold source temperature (273 K). Here, the total entropy generation is equal to 26.00 W/K. Furthermore, the COA of this heat pump was found to be equal to 3.58 (see section 2.2.3.2 for the level of complexity 3). The main quantities relating to the analysis in terms of entropy generation number are presented in Table 2.8. As announced theoretically, and even in the case of heat rejection or gain, the thermodynamic efficiency is eventually linked to the entropy generation number by the relation $\eta = 1 - N_s$.

Refrigerator	Heat pump
$COP = 2.43$	$COA = 3.58$
$COA_{Carnot} = \dfrac{273}{303 - 273} = 9.1$	$COA_{Carnot} = \dfrac{303}{303 - 273} = 10.1$
$\eta = \dfrac{COP}{COP_{Carnot}} = 0.300$	$\eta = \dfrac{COA}{COA_{Carnot}} = 0.354$
$P(s)_{max} = \dfrac{\dot{W}}{T_c} = 36.03\ W/K$	$P(s)_{max} = \dfrac{\dot{W}}{T_f} = 40.29\ \mathrm{W/K}$
$N_s = P(s)/P(s)_{max} = \dfrac{25.45}{36.03} = 0.700$	$N_s = P(s)/P(s)_{max} = \dfrac{26.00}{40.29} = 0.646$
$\eta = 1 - N_s = 0.300$	$\eta = 1 - N_s = 0.354$

Table 2.8. *Main quantities relating to the analysis in terms of entropy generation number for the reverse cycle engine studied, operating as a refrigeration system or as a heat pump*

2.3.4. *Phenomenon-scale entropy analysis: two-phase flows with heat transfer and phase change*

The component-scale entropy analysis applied to the evaporator and condenser remains unsatisfactory. Indeed, by reasoning on the internal equivalent source temperatures, the irreversibilities created in the heat exchanger are not quantified, but reasoning on the external source does not make it possible to evaluate the performance of the refrigeration circuit itself, which nevertheless forms the heart of the system that the engineer is looking to design. It is therefore the entropy analysis of the phenomena, i.e. along the fluid flows in the exchangers, which will make it possible to reveal the phenomena that create entropy and to consider optimization strategies. The entropy analysis of heat transfer by single-phase convection (cooled fluid or secondary refrigerant in the evaporator, or heat transfer fluids in the condenser) is relatively classic, as is the entropy analysis of heat transfer by conduction in solid elements that constitute the walls of the exchangers. This section therefore deals with the entropy analysis of two-phase flows within which phase change heat transfers occur.

2.3.4.1. *Phase-separated two-phase flow model*

Consider the flow of a two-phase liquid–vapor fluid at the saturation temperature T_{sat} in a tube with a linear fluid contact surface Σ (unit: m²/m; for a smooth circular tube of diameter d, where $\Sigma = \pi d$). Imagine an elementary volume of the tube of length dz at temperature T_p exchanging an elementary thermal power with the fluid $\delta\dot{Q}$. The total fluid flow ($\dot{m}$) is composed of a flow of liquid ($\dot{m}_l$) of entropy s_l and a vapor flow ($\dot{m}_v$) of entropy s_v. Under the effect of heat exchange (heat input for an evaporator, heat extraction for a condenser) and pressure drop which leads to an evaporation phenomenon (the term "flashing" is commonly used to designate this phenomenon), the vapor content of the two-phase mixture ($x = \frac{\dot{m}_v}{\dot{m}_v+\dot{m}_l}$) varies by an amount dx in the elementary volume. The second law of thermodynamics applied to this tube element makes it possible to express the entropy generation:

$$d\dot{s}_c = d(\dot{m}_v s_v + \dot{m}_l s_l) - \frac{\delta\dot{Q}}{T_p} \qquad [2.66]$$

Here, $\dot{m} = \dot{m}_l + \dot{m}_v$ and by introducing the elementary entropy variations of the two-phase mixture ($ds_{tp} = d(\dot{m}_v s_v + \dot{m}_l s_l)/\dot{m}$) and enthalpy of the two-phase mixture $dh_{tp} = T_{sat} ds_{tp} + v_{tp} dp$, where p represents the pressure in the tube element and v_{tp} is the specific volume of the two-phase mixture which can be calculated from the mixture vapor content in the elementary volume and the specific volumes of each phase (v_l and v_v) by a simple weighting ($v_{tp} = (1-x)v_l + xv_v$), the previous equation can be rewritten as follows:

$$d\dot{s_c} = \frac{\dot{m}}{T_p} dh_{tp} - \frac{\delta \dot{Q}}{T_p} - \frac{\dot{m}v_{tp}}{T_p} dp \quad [2.67]$$

The first law of thermodynamics applied to the elementary volume of a tube is written as:

$$\frac{\delta \dot{Q}}{\dot{m}} = dh_{tp} = h_{lv}dx + xdh_v + (1-x)dh_l \quad [2.68]$$

where h_{lv} is the latent heat of vaporization at the considered saturation temperature. The temperature difference between the wall and the fluid is introduced here $\Delta T = T_p - T_{sat}$ ($\Delta T > 0$ in the evaporator but $\Delta T < 0$ in the condenser) to express the elementary local entropy generation associated with two-phase flow in the presence of heat exchange and phase change:

$$\frac{d\dot{s_c}}{dz} = \frac{\delta \dot{Q}}{dz}\left[\frac{\Delta T}{(1+\frac{\Delta T}{T_{sat}})T_{sat}^2}\right] + \frac{\dot{m}v_{tp}}{T_{sat}}\left(-\frac{dp}{dz}\right) \quad [2.69]$$

It is often preferred to reason in terms of heat flux density and introduce the heat transfer coefficient:

$$q = h\Delta T = \frac{\delta \dot{Q}}{dz.\Sigma} \quad [2.70]$$

Here, $\Delta T/T_{sat} \ll 1$, with the local entropy generation being determined using this expression:

$$\frac{d\dot{s_c}}{dz} = \frac{q^2\Sigma}{hT_{sat}^2} + \frac{\dot{m}v_{tp}}{T_{sat}}\left(-\frac{dp}{dz}\right) \quad [2.71]$$

Integrated over the entire length of the exchanger circuit, the local elementary entropy generation makes it possible to determine the entropy generation associated with the two-phase flow with heat transfer in the exchanger. In addition, this expression makes it possible to highlight the contribution to the overall entropy generation of the heat transfer between the wall and the fluid (first term of the right-hand side) and of the pressure drops (second term of the right-hand side). The entropy generation is reduced when the heat transfer is enhanced (increase in the heat transfer coefficient) and when the pressure drops are reduced. This also reveals that to improve two-phase heat exchangers (i.e. to reduce the irreversibilities created within them), it is essential to have reliable tools to evaluate the heat transfer coefficients and the pressure drops, two subjects which must therefore remain at the heart of the research in the field of reverse cycle engines.

2.3.4.2. *Mixing model of a phase-separated two-phase flow*

The overall two-phase flow model presented in the above section can be criticized in the sense that it disregards a common but difficult to quantify phenomenon, namely the velocity slip. In fact, a higher vapor velocity is frequently observed (U_v) than the liquid velocity (U_l), a phenomenon partly due to a vapor density (ρ_v) significantly lower than that of the liquid (ρ_l). If the slip ratio $\Gamma = U_v/U_l$ deviates sufficiently from 1, the thermodynamic vapor quality defined previously from the flow rates ($x = \frac{\dot{m}_v}{\dot{m}_v+\dot{m}_l}$) is not representative of the ratio of the masses of vapor and fluid contained in the elementary volume of the tube, which will be noted $x_m = \frac{m_v}{m_v+m_l}$. The relationship between this vapor content of the mixture, the vapor and liquid flow rates and the slip ratio is expressed as follows:

$$x_m = \frac{\dot{m}_v}{\dot{m}_v+\Gamma\dot{m}_l} = \frac{x}{x+\Gamma(1-x)} \qquad [2.72]$$

To take into account the phenomenon of velocity slip in the calculation of the local entropy generation, the following expression is used[1]:

$$\frac{d\dot{s_c}}{dz} = \frac{q^2\Sigma}{FhT_{sat}^2} + \frac{\dot{m}v_m}{T_{sat}}\left(-\frac{dp}{dz}\right) \qquad [2.73]$$

with :

$$v_{tp} = (1-x)v_l + xv_v$$

$$v_m = (1-x_m)v_l + x_m v_v \qquad [2.74]$$

$$F = \frac{\Gamma h_{lv}\frac{\delta\dot{Q}}{dz}}{\beta\Gamma h_{lv}\cdot\frac{\delta\dot{Q}}{dz} - [(c_{x,m}-\beta c_{x,tp})(v_v-v_l)T_{sat}+h_{lv}(v_m-\beta v_{tp})]\Gamma\dot{m}\left(-\frac{dp}{dz}\right) - x(1-x)\dot{m}h_{lv}^2\beta\frac{d\Gamma}{dz}} \qquad [2.75]$$

$$\beta = \frac{\Gamma}{(x+\Gamma(1-x))^2} \qquad [2.76]$$

$$c_{x,m} = (1-x_m)c_{x,l} + x_m c_{x,v} \qquad [2.77]$$

1. The, relatively heavy, mathematical developments making it possible to obtain this expression were presented by Revellin et al. (2009).

$$c_{x,l} = T_{sat} \left(\frac{ds_l}{dT_{sat}}\right)_{x=0} \quad [2.78]$$

$$c_{x,v} = T_{sat} \left(\frac{ds_v}{dT_{sat}}\right)_{x=1} \quad [2.79]$$

The variables $c_{x,v}$ and $c_{x,l}$ are specific heat capacities at constant quality, and are therefore characteristic thermodynamic properties of the refrigerant – just as are specific heat capacities at constant pressure (c_p) or at constant volume (c_v) which are in much more widespread use. They can be evaluated from fluid property tables or state equations.

This phase-separated flow model for the calculation of the entropy generation is more complex than the overall model presented previously, but it is more complete since it includes the effect of the velocity slip, which is at the origin of the irreversibilities and increases as the slip ratio increases. Its use is completely comparable to that of the overall model: the integration of local elementary entropy generation (equation [2.73]) along the circuit of the exchanger makes it possible to determine the entropy generation associated with the two-phase flow with heat transfer. It also makes it possible to study the entropy generation distribution, to consider improved heat exchanger designs (i.e. making it possible to limit irreversibilities and to evenly distribute them, to minimize them) and thus to increase the performance reverse cycle. Note that this approach requires, in addition to the knowledge of the pressure drop and the local transfer coefficient – as they were required to implement the overall model, also knowledge of the local velocity slip ratio. It is also an area of research in laboratories contributing to the development of the understanding of phase change two-phase flows, often approached from the angle of the development of tools for the determination of the void fraction $\varepsilon = \frac{A_v}{A_v + A_l}$ (ratio of the cross-section of the line occupied by the vapor to the total cross-section of the line along the exchangers, depending on the flow conditions). Indeed, the slip ratio and the void fraction are linked to the vapor quality by the following relationship:

$$\varepsilon = \left[1 + \Gamma\left(\frac{1-x}{x}\right)\left(\frac{v_l}{v_v}\right)\right]^{-1} \quad [2.80]$$

The determination of the vapor quality along the circuit of the exchanger (using heat and momentum balances) and the determination of the void fraction using such tools make it possible to determine the slip ratio and finally to implement the phase-separated flow model for the entropy analysis of two-phase flows with phase change heat transfer.

2.4. Exergy analysis

2.4.1. *From the concept of exergy to proposed definitions*

The concept of exergy was coined by Georges Gouy who, at the beginning of the 20th century, defined the concept of "usable energy" (Gouy 1889), which is now called exergy. If a system interacts with an environment which constitutes an infinite source (a reservoir) at constant temperature and pressure, and of fixed composition, the maximum usable energy resulting from exchanges between the system and this reservoir depends on the difference between the temperatures, pressures and compositions. In the case of heat exchanges, the maximum usable energy therefore depends on the temperature of the system and the temperature of the environment. This observation echoes Lord Kelvin's expression that "heat is a degraded form of energy". This expression sums up the observation that mechanical work can easily be converted entirely into heat, but heat cannot be entirely converted into work: the usable energy of a quantity of heat depends on its temperature and on the temperature of its environment.

The concept has since flourished and other expressions have been proposed. For example, Professor Daniel Favrat defines exergy as follows: "The exergy associated with a transfer or a stock of energy is defined as the maximum work potential that it would ideally be possible to draw from each energy unit transferred or stored (using reversible cycles with the atmosphere as one of two sources – cold or hot)." A quantity of heat available at the temperature of the environment has no work potential, but the greater the difference between the temperature of the environment and the temperature at which the quantity of heat available is high, the greater the work potential.

For Professors Riad Benelmir, André Lallemand and Michel Feidt, "the exergy of a certain quantity of matter contained in a system is a measure of the potential for generation (or reception) of maximum (or minimum) work by the supersystem (consisting of the system and its surrounding medium), which will allow this quantity of matter to be brought back from its initial state to a state of inert equilibrium with the surrounding medium" (Benelmir et al. 2002).

From these definition proposals, any quantity of energy (respectively: any energy power) can be associated with an exergy (respectively: exergy power). If the energy considered is noble (mechanical, electrical work, etc.), then its exergy represents the totality of this energy. If this energy is a degraded, non-noble form (heat), then its exergy is only a fraction of this thermal energy. However, these definitions are marked by a vision particularly oriented towards thermodynamic systems of thermo-mechanical conversion, i.e. the generation of mechanical power from heat sources, for which exergy analysis has become relatively classic. The extension of exergy theories to reverse cycle engines raises specific questions, which will be discussed in the next sections. It is first of all from the energy and entropy balances that a double mathematical definition of exergy can be formulated.

2.4.2. *Mathematical definitions of exergy*

2.4.2.1. *Exergy content and rate*

To propose a mathematical definition of the exergy rate associated with a thermal power $\dot{Q}$ at temperature T, taking into account the proposed definitions, Gouy, Favrat or even Benelmir, Feidt and Lallemand consider the use of a Carnot engine (reversible engine) to convert this thermal power into mechanical power, a form of noble energy. Such a Carnot engine can only operate if it has a heat source capable of absorbing the heat rejection from the engine (thermal power $\dot{Q}_{loss}$). For the sake of generality, this source is assumed to be at a reference temperature (T_{ref}), even if usage often leads us in terms of engines (even for a fictitious engine like the Carnot engine, because it represents the reversible upper limit of real engines to which the latter will be compared) to choose as a source the atmosphere at the temperature T_{amb}. This Carnot engine is shown in Figure 2.10, which shows the entropic power $\dot{S}$ transferred from the hot source of the engine (temperature T) to the engine itself, and this same rate of entropy released from the engine to the cold source at the temperature T_{ref}.

The efficiency $\eta = -\dot{W}/\dot{Q}$ of such a Carnot engine is equal to $1 - T_{ref}/T$. This result, which is very typical, can be found easily by combining the first and second laws of thermodynamics applied to the reversible engine (i.e. for a zero-entropy generation) in steady state:

$$\dot{W} + \dot{Q}_{loss} + \dot{Q} = 0 \qquad [2.81]$$

$$\frac{\dot{Q}_{loss}}{T_{ref}} + \frac{\dot{Q}}{T} = 0 \qquad [2.82]$$

For a given thermal power $\dot{Q}$, any entropy generated in the engine (which would then lose its "Carnot" character) would be accompanied by an increase in heat rejection and consequently a decrease in the mechanical work produced. Thus, the maximum power produced by such an engine is that produced by the Carnot engine. However, this maximum power is by definition the exergy rate associated with the thermal power $\dot{Q}$ at temperature T, which is written as:

$$\dot{Ex}_{\dot{Q},T} = \left(1 - \frac{T_{ref}}{T}\right)\dot{Q} \qquad [2.83]$$

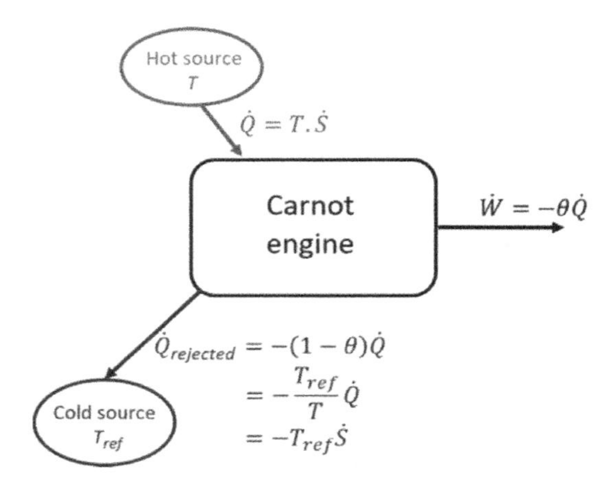

Figure 2.10. *Carnot engine*

In this context, the "Carnot factor" is the multiplier term of the thermal power:

$$\theta = \left(1 - \frac{T_{ref}}{\mathrm{T}}\right) \quad [2.84]$$

In accordance with previous reflections, the exergy rate associated with the mechanical power $\dot{W}$ is entirely this mechanical power:

$$\dot{Ex}_{\dot{W}} = \dot{W} \quad [2.85]$$

We can also approach the concept of the exergy rate by means of a linear combination of the first and second laws of thermodynamics applied to any closed system: from such energy and entropy balances, we can formulate an exergy balance, including an unsteady term if necessary. The linear combination of equations [2.1] and [2.30], multiplying each term of the second law by the opposite of the reference temperature, leads to:

$$\frac{dE}{dt} - T_{ref}\frac{dS}{dt} = \sum\left(\dot{W}_i + \left(1 - \frac{T_{ref}}{T_i}\right).\dot{Q}_i\right) - T_{ref}.P(s) \quad [2.86]$$

This combination reveals an exergy accumulation term (left-hand side), the exergy rates associated with the mechanical and thermal powers exchanged by the system, and a term called "exergy destruction rate" ($\dot{Ex}_d = T_{ref}.P(s)$)

$$\frac{dEx}{dt} = \sum \dot{Ex}_{\dot{W}_i} + \sum \dot{Ex}_{\dot{Q}_i,T_i} - \dot{Ex}_d \quad [2.87]$$

Thus, since the irreversibilities appear to be a source of entropy generation, they can also be interpreted as exergy destruction. In the limit, a reversibly functioning system is therefore characterized by the absence of exergy destruction.

Finally, it should be emphasized that the considerations developed here in terms of power (mechanical power, thermal power and associated exergy rates) remain valid in terms of energy. The exergy content of mechanical work is this work itself, and the exergy associated with a thermal energy depends on the temperature level and is equal to the Carnot factor multiplied by this thermal energy. Finally, the product of the reference temperature by the creation of entropy is the destroyed exergy.

2.4.2.2. *Exergy as a state function*

The reasoning conducted in the previous section (see section 2.4.2.1) for a closed system can be extended to an open system, for example each component of a reverse cycle engine, operating in steady state. From the same method of linear combination as before (i.e. by carrying out a linear combination of equations [2.2] and [2.31], by multiplying each term of the second law by the opposite of the reference temperature), the following can be written as:

$$0 = \sum \dot{Ex}_{\dot{W}_i} + \sum \dot{Ex}_{\dot{Q}_i,T_i} + \sum \left(\dot{m}_e(h_e - T_{ref}s_e) - \dot{m}_s(h_s - T_{ref}s_s)\right) - \dot{Ex}_d \quad [2.88]$$

This equation reveals an exergy flux term, which leads us to define the specific exergy of the fluid as:

$$ex = h - T_{ref}s \quad [2.89]$$

Note that if the kinetic energy and the gravitational potential energy intervene in the analysis, the specific exergy is based on the total enthalpy $h_t = h + e_c + e_p$. Furthermore, apart from the case of cycled fluids or in the event of a variation in the volume of the system, a term comes into play, comprising a reference pressure, which represents the pressure of the environment of the system, like the reference temperature which generally represents the environment temperature. It does not seem useful to develop these considerations in the context of reverse cycle engines.

2.4.3. *Exergy analysis of reverse cycle engines*

To illustrate the exergy analysis of a reverse cycle engine, the system designed at complexity level 3 is of most interest (Figure 2.2 and Table 2.5). The application of the exergy balance (equation [2.87]) in steady state leads to:

$$0 = \dot{E}x_{\dot{W}} + \dot{E}x_{\dot{Q}_{cond},T_c} + \dot{E}x_{\dot{Q}_{evap},T_f} + \dot{E}x_{\dot{Q}_{comp},T_{amb}} + \dot{E}x_{\dot{Q}_{ref},T_{amb}} + \dot{E}x_{\dot{Q}_{asp},T_{amb}} - \dot{E}x_d \quad [2.90]$$

It is at this stage that the choice of the reference temperature comes into play, which can be guided by different points of view, in particular depending on whether the engine is used as a refrigerator or a heat pump.

2.4.3.1. *Refrigerator*

For a refrigerator, the cold source temperature is lower than the ambient temperature ($T_f < T_{amb}$); otherwise, the use of an engine is absurd since the atmosphere could be used to produce a cooling effect. On the other hand, the temperature of the hot source can be, but is not necessarily equal to, the ambient temperature.

In the case where the heat source is the ambient air (i.e. $T_c = T_{amb}$), a choice is practically imposed on the analyst: it seems more than sensible to adopt in this case $T_{ref} = T_c = T_{amb}$. Indeed, the exergy balance then reduces to:

$$\dot{E}x_{\dot{W}} + \dot{E}x_{\dot{Q}_{evap},T_f} - \dot{E}x_d = 0 \quad [2.91]$$

and so

$$\dot{W} + \dot{Q}_{evap}\left(1 - \frac{T_{ref}}{T_f}\right) - T_{ref}P(s) = 0 \quad [2.92]$$

Note that the exergy rate associated with the cooling capacity $\dot{E}x_{\dot{Q}_{evap},T_f}$ is negative due to the Carnot factor sign $\theta_f = \left(1 - \frac{T_{ref}}{T_f}\right)$. The exergy balance expresses that the exergetic power (exergy rate) supplied to the system ($\dot{E}x_{\dot{W}} = \dot{W}$) is partly converted into the cooling exergy rate and partly destroyed due to the irreversible nature of the thermodynamic transformations.

It is then easy to define the exergy efficiency of the system:

$$\eta_{ex} = \frac{-\dot{E}x_{\dot{Q}_{evap},T_f}}{\dot{E}x_{\dot{W}}} \quad [2.93]$$

which is expressed as

$$\eta_{ex} = \frac{\dot{Q}_{evap}\left(\frac{T_{ref}}{T_f} - 1\right)}{\dot{W}} \quad [2.94]$$

Given the fact that $T_{ref} = T_c$ and recalling the definition of the COP and the expression of the Carnot COP based on the source temperatures (equation [2.35]), note that:

$$\eta_{ex} = \frac{COP}{COP_{Carnot}} \qquad [2.95]$$

It must be emphasized that this simple expression is valid only if the reference temperature of the exergies is equal to the temperature of the heat released. When several sources at different temperatures are involved (e.g. if the compression is cooled by means of a circulation of water taken from a river while the hot source is the ambient air), the choice of the reference temperature is not necessarily obvious and may invalidate the proposed expression. The fact remains that the exergy efficiency, calculated using equation [2.94], will provide an indication of the deviation from reversibility, while the exergy balance (equation [2.90]) will highlight the relative importance of heat rejection expressed in the exergy rate. Such a situation will be illustrated numerically in a case study (section 2.5.1).

On the other hand, if the heat rejection temperature is adopted as a reference temperature, then the exergy efficiency constitutes an immediate tool to quantify the deviation of the system from ideality (i.e. from reversible operation).

2.4.3.2. *Heat pump*

For a heat pump, the hot source temperature is higher than the ambient temperature ($T_c > T_{amb}$); otherwise, the use of an engine is absurd. As for the refrigerator, a simple form of the exergy balance and the exergy efficiency is obtained if the chosen reference temperature is such that $T_{ref} = T_f = T_{amb}$. In this case, the exergy balance is reduced to:

$$\dot{Ex}_{\dot{W}} + \dot{Ex}_{\dot{Q}_{cond},T_c} - \dot{Ex}_d = 0 \qquad [2.96]$$

And so

$$\dot{W} + \dot{Q}_{cond}\left(1 - \frac{T_{ref}}{T_c}\right) - T_{ref}P(s) = 0 \qquad [2.97]$$

In these cases, the exergy rate associated with the heat generation $\dot{Ex}_{\dot{Q}_{cond},T_c}$ is negative because the thermal power produced $\dot{Q}_{cond}$ is negative. But as for the refrigerator, the exergy balance indicates directly that the exergetic power (exergy rate) supplied to the system ($\dot{Ex}_{\dot{W}} = \dot{W}$) is partly converted into exergy heating power and the rest is destroyed in the form of irreversibilities.

It is then easily shown that under these conditions, the exergy efficiency:

$$\eta_{ex} = \frac{-\dot{Ex}_{\dot{Q}_{cond},T_C}}{\dot{Ex}_{\dot{W}}} = \frac{-\dot{Q}_{cond}\left(1-\frac{T_{ref}}{T_C}\right)}{\dot{W}} \quad [2.98]$$

can be simplified like this:

$$\eta_{ex} = \frac{COA}{COA_{Carnot}} \quad [2.99]$$

As in the case of a refrigerator, this expression cannot be considered as a generality. For example, it is common for the compressor and the evaporator to be located in different environments and thus exchange with different sources (e.g. buried evaporator of a geothermal heat pump and the compressor in the room to be heated or in a technical zone), for which no choice of reference temperature seems obvious and will not lead to a simple relationship like the previous equation. On the other hand, and here again analogously to the refrigerator, if the reference temperature is the temperature of the cold source, the exergy efficiency is essential in the sense that it directly indicates the deviation from the ideality of the studied engine.

2.5. Case study for exergy analysis

2.5.1. *Refrigerator with cooled compression and recovery of heat rejected*

On an industrial food-processing site, a refrigerator is used to produce a cooling power $\dot{Q}_f = 25$ kW at the temperature of the cold source $T_f = -10$°C. The condensation heat $\dot{Q}_c = -29.5$ kW is released towards the ambient air at the temperature $T_c = 30$°C. The compressor absorbs a power equal to $\dot{W} = 9.5$ kW and, due to technological constraints (limitation of the discharge temperature), it must be cooled by circulating water at the temperature $T_{water} = 12$°C. The cooling circuit makes it possible to evacuate a thermal power of $\dot{Q}_{water} = -5.5$ kW. The suction line is the site of a heat input $\dot{Q}_{in} = 0.5$ kW from its environment at the temperature $T_{in} = 0$°C. Finally, the operation of the site requires a heating power $\dot{Q}_{calo} = 10$ kW at temperature $T_{calo} = 40$°C. This power is supplied by means of electrical resistors ($|\dot{W}_{calo}| = |\dot{Q}_{calo}|$). The heat rejection temperature is chosen as the reference temperature for the exergies: $T_{ref} = T_c = 303$ K.

The energy balance of the refrigerator makes it possible to check that the exchanges of heat and mechanical energies have been correctly recorded, since

$\dot{Q}_c + \dot{Q}_f + \dot{Q}_{\text{in}} + \dot{Q}_{water} + \dot{W} = -29.5 + 25 + 0.5 - 5.5 + 9.5 = 0$. The exergy balance makes it possible to determine the rate of exergy destruction:

$$\dot{Ex}_d = \dot{Ex}_{\dot{W}} + \dot{Ex}_{\dot{Q}_c,T_c} + \dot{Ex}_{\dot{Q}_f,T_f} + \dot{Ex}_{\dot{Q}_{water},T_{water}} + \dot{Ex}_{\dot{Q}_{in},T_{in}} \qquad [2.100]$$

where $\dot{Ex}_d = 9.5 + 0 - 3.80 + 0.34 - 0.05 = 5.99$ kW. Note the negative sign of $\dot{Ex}_{\dot{Q}_f,T_f}$ (cooling capacity, see section 2.3.4.1), the positive sign of $\dot{Ex}_{\dot{Q}_{water},T_{water}}$ (heat rejected towards a "cold" source, i.e. at a temperature lower than the reference temperature) and the negative sign of $\dot{Ex}_{\dot{Q}_{in},T_{in}}$ (heat input from a "cold" source, i.e. also at a lower temperature than the reference temperature). The COP of the refrigerator $COP = \dot{Q}_f/\dot{W} = 2.63$ can be related to the Carnot COP $COP_{Carnot} = T_f/(T_c - T_f) = 6.575$. The exergy efficiency $\eta_{ex} = -\dot{Ex}_{\dot{Q}_f,T_f}/\dot{Ex}_{\dot{W}} = 3.80/9.5 = 0.4$ is equal to the ratio COP/COP_{Carnot}.

At the site scale, an exergy efficiency can be defined and calculated taking into account the electricity consumption to cover the heating need:

$$\eta_{ex} = \frac{-\dot{Ex}_{\dot{Q}_f,T_f} + \dot{Ex}_{\dot{Q}_{calo}}}{\dot{W} + \dot{W}_{calo}} = \frac{\dot{Q}_f\left(\frac{T_{ref}}{T_f} - 1\right) + \dot{Q}_{calo}\left(1 - \frac{T_{ref}}{T_{calo}}\right)}{\dot{W} + \dot{W}_{calo}} = 0{,}211 \qquad [2.101]$$

To increase the site's exergy efficiency, a different setting is made for the refrigerator, so as to use part of the heat rejected to the condenser to replace it with the heat supplied by the electrical resistors. This new setting leads to the following modified values: $\dot{W}^* = 11.5$ kW (the increase in the discharge temperature requires a higher condensing temperature, and finally an increase in the compression ratio which results in a higher absorbed power), $\dot{Q}^*_{water} = -7.6$ kW (with the compressor operating at a higher compression ratio, more thermal power must be removed to maintain an acceptable discharge temperature), $\dot{Q}^*_c = -19.4$ kW. The energy balance respects the first law of thermodynamics: $\dot{Q}^*_c + \dot{Q}_f + \dot{Q}_{\text{in}} + \dot{Q}^*_{water} + \dot{W}^* - \dot{Q}_{calo} = -19.4 + 25 + 0.5 - 7.6 + 11.5 - 10 = 0$.

At the site scale, the exergy efficiency is then expressed as:

$$\eta_{ex} = \frac{-\dot{Ex}_{\dot{Q}_f,T_f} - \dot{Ex}_{\dot{Q}_{calo}}}{\dot{W}} = \frac{\dot{Q}_f\left(\frac{T_{ref}}{T_f} - 1\right) - \dot{Q}_{calo}\left(1 - \frac{T_{ref}}{T_{calo}}\right)}{\dot{W}} = 0.358 \qquad [2.102]$$

The gain in terms of exergy efficiency is particularly clear. It comes from the fact that the electricity (pure exergy) used to produce the heating power at a temperature of 40°C, barely higher than the reference temperature (very low exergy content), was a real exergy waste, usefully replaced by the recovery of heat, which would otherwise constitute heat rejection. This example demonstrates the interest of conducting exergy analyses when several temperature levels coexist and it is possible to carry out operations to recover hot or cold flows.

2.5.2. *Heat pump running on CO_2 with or without an ejector*

Carbon dioxide is considered a useful refrigerant for its application in heat pumps. It is non-toxic and non-flammable, and has good compatibility with many oils, low cost and favorable heat transfer properties. Nevertheless, CO_2 exhibits large irreversibilities during its expansion and during its supercritical cooling as it dissipates heat towards the hot source. The cooling is said to be supercritical in the sense that in general, when using heat pumps, the hot source temperature is higher than the critical temperature of the CO_2 ($T_{crit} = 31.6°C$), so that the fluid remains in the supercritical state throughout the thermodynamic transformation which accompanies the heat exchange towards the hot source. The use of an ejector can make it possible to compensate for the disadvantage induced by the irreversibilities associated with expansion. The gain will be studied in this case study.

Consider a transcritical semi-instantaneous thermodynamic water heater operating on CO_2 (Figure 2.11) which heats domestic hot water (DHW) from 15 to 60°C for three apartments. This system consists of a gas cooler which dissipates heat towards the hot source ($\dot{Q}_{hot}$), an evaporator ($T_{evap} = -2°C$) exchanging with a cold source (external air at the temperature $T_f = 6°C$) a power of $\dot{Q}_{evap} = 3.28$ kW, a compressor (whose heat rejections are assumed to be low enough for it to be considered adiabatic) which absorbs a power of $\dot{W} = 1.82$ kW, an adiabatic separator and an intermediate exchanger. In addition, in order to improve its operation, this engine has an ejector, which makes it possible to recover part of the energy from expansion to compress the fluid leaving the evaporator and thus reduce the consumption of the compressor.

By applying the first law in steady state, the thermal power exchanged with the hot source (DHW) can thus be calculated as:

$$\dot{Q}_{hot} = -\dot{W} - \dot{Q}_{evap} = -1.82 - 3.28 = -5.10 \text{ kW}$$

The COA of this installation is therefore $\text{COA} = \frac{-\dot{Q}_{hot}}{\dot{W}} = \frac{5.10}{1.82} = 2.80$.

This COA value may seem satisfactory, but the results of the exergy analysis, in particular the calculation of the exergy destruction power and the exergy efficiency, moderate this impression.

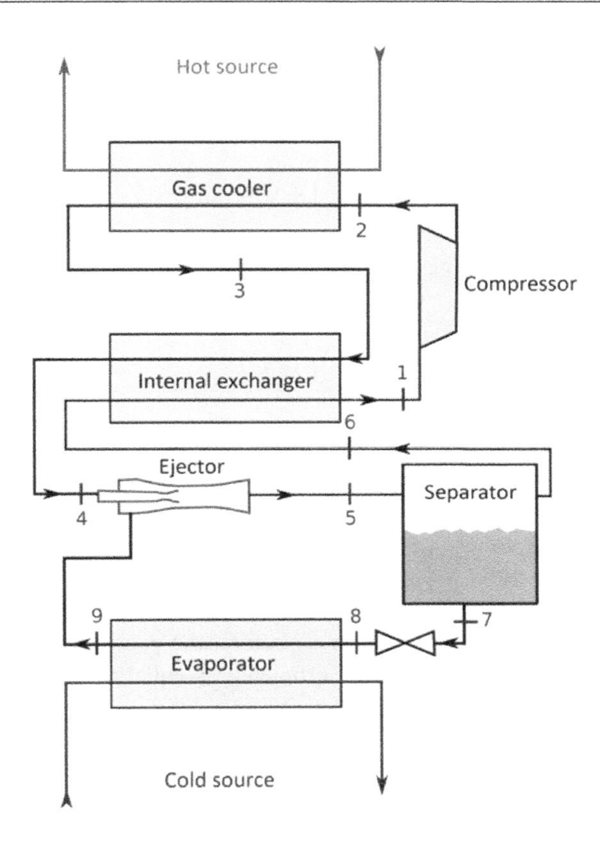

Figure 2.11. *Diagram of an ejection heat pump*

Since the hot source is not at a uniform temperature (since the water heats up from T_{inlet} = 15°C to T_{outlet} = 60°C), the entropic temperature of the water is adopted as the equivalent source temperature. For a single-phase liquid of constant specific heat capacity, the entropic temperature (equation [2.40]) can be calculated as the logarithmic average of inlet and outlet temperatures: $\tilde{T}_c = \frac{T_{inlet}-T_{outlet}}{\ln(\frac{T_{inlet}}{T_{outlet}})}$.

Therefore, $\tilde{T}_c = \frac{(15+273)-(60+273)}{\ln(\frac{15+273}{60+273})} = 310.0$ K. If the difference between the inlet and outlet temperatures remains modest, the entropic temperature can be approximated by the arithmetic mean.

The rate of exergy destruction can be deduced from the exergy balance applied to the complete system: $-\dot{E}x_d = \dot{W} + \dot{Q}_{hot}\left(1 - \frac{T_{ref}}{\tilde{T}_c}\right) + \dot{Q}_{evap}\left(1 - \frac{T_{ref}}{T_f}\right)$.

In accordance with the practice suggested in section 2.4.3.2, the reference temperature $T_{ref} = T_f = 279$ K is chosen, so that the exergy rate associated with the heat exchange at the cold source is zero. It becomes:

$$-\dot{E}x_d = 1.82 - 5.10\left(1 - \frac{279}{310.0}\right) + 0 = 1.31 \text{ kW}$$

Finally, the exergy efficiency of the heat pump is:

$$\eta_{ex} = \frac{\dot{Q}_{hot}\left(1 - \frac{T_{ref}}{\tilde{T}_c}\right)}{\dot{W}} = \frac{0.51}{1.82} = 0.280$$

Imagining that this heat pump was designed without an ejector, for the same heating power ($\dot{Q}_{hot} = -5.10$ kW), it can be shown that, taking into account the modification of the cycle and the flow involved, the thermal power extracted at the cold source is equal to $\dot{Q}_{evap} = 2.91$ kW. This means that the power of the compressor is equal to $\dot{W} = -\dot{Q}_{hot} - \dot{Q}_{evap} = 2.19$ kW.

By performing the same calculations as before, the exergy destruction power is equal to $-\dot{E}x_d = 1.68$ kW, the exergy efficiency is equal to $\eta_{ex} = 0.233$ and the amplification coefficient is equal to $COA = 2.33$.

From this simple case study:

– the ejector makes it possible to increase by 20% the COA as well as the exergy efficiency;

– the exergy efficiency of the system with ejector remains a little low (close to but less than 0.3) compared to values that can be achieved with competing systems. This tends to indicate that there is still room for improvement in this system.

2.6. References

Bejan, A. (1982). *Entropy Generation through Heat and Fluid Flow*. Wiley, New York.

Benelmir, R., Lallemand, A., Feidt, M. (2002). Analyse exergétique. *Techniques de l'Ingénieur*. BE 8015.

Gouy, G. (1889). Sur l'énergie utilisable. *Journal de physique théorique et appliquée*, 8(1), 501–518.

Revellin, R., Lips, S., Khandekar, S., Bonjour, J. (2009). Local entropy generation for saturated two-phase flow. *Energy*, 34, 1113–1121. doi:10.1016/j.energy.2009.03.014.

3

Thermodynamics and Optimization of Reverse Cycle Engines

Michel FEIDT
Université de Lorraine, LEMTA, CNRS, Nancy, France

3.1. Reverse cycle engines according to equilibrium thermodynamics: reminders of the concepts

The first models of reverse cycle engines were developed on the basis of equilibrium thermodynamics. To illustrate the fundamental concepts, we consider the example of a two-heat-source engine operating according to a reverse Carnot cycle (Figure 3.1).

The cold source is an infinite heat source at constant temperature T_{CS}; the hot sink is an infinite heat source at temperature T_{HS}. The source and the sink are therefore thermostats (hypothesis 1) (Carnot 1824).

The thermodynamic system under study consists of a reverse cycle engine represented by the 1-2-3-4 cycle and the two thermostats.

Assuming no heat is exchanged by the system but with the sources (hypothesis 2), the cycled fluid receives the thermal energy Q_{CS} at the cold end and delivers the

---

For a color version of all the figures in this chapter, see www.iste.co.uk/bonjour/refrigerators.zip.

Refrigerators, Heat Pumps and Reverse Cycle Engines,
coordinated by Jocelyn BONJOUR.

thermal energy Q_{HS} at the hot end. These heat exchanges require the mechanical energy W such that the energy balance over the cycle is written as:

$$W = Q_{HS} - Q_{CS} \quad [3.1]$$

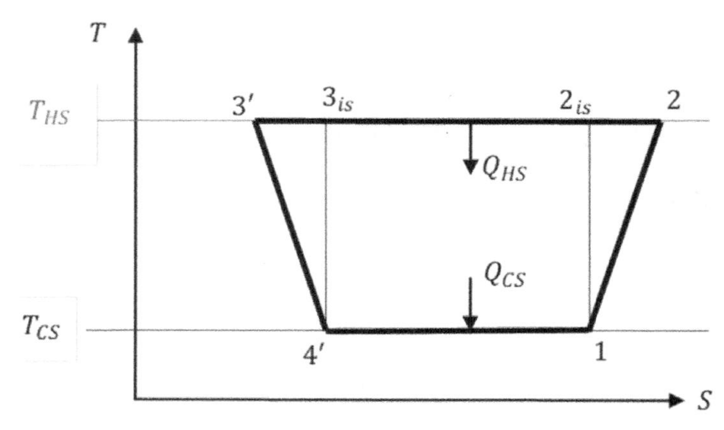

Figure 3.1. *Reverse endoreversible or endo-irreversible Carnot cycle (reference to the cold end)*

The actual engine is subject to irreversible transformations, according to the 1-2-3-4 cycle, leading to the entropy balance of the cycle:

$$\frac{Q_{HS}}{T_{HS}} = \frac{Q_{CS}}{T_{CS}} + \Delta SI \quad [3.2]$$

Here, ΔSI represents the entropy generation associated with the 1-2-3-4 cycle. In the case of the reversible cycle, $\Delta SI = 0$, so that the combination of relations [3.1] and [3.2] leads to an efficiency limit value of the engine, in the sense of the first law, in accordance with the 1-2_{is}-3_{is}-4 cycle (Figure 3.1). This energy efficiency is generally defined by the ratio of the useful heating energy (UHE) to the mechanical energy expenditure (MEE) necessary for generating the useful effect:

$$z_i = \frac{UCE}{MEE} \quad [3.3]$$

Thus, for the refrigerator, the coefficient of performance COP_{RM} becomes:

$$COP_{RM_{rev}} = \frac{Q_{CS}}{W} = \frac{T_{CS}}{T_{HS} - T_{CS}} \quad [3.4]$$

and for the heat pump,

$$COP_{HPrev} = \frac{Q_{HS}}{W} = \frac{T_{HS}}{T_{HS}-T_{CS}} \quad [3.5]$$

Note in relations [3.4.] and [3.5] that these efficiencies can be greater than 1, unlike the efficiency of the engines. Also note the relationship between these two coefficients:

$$COP_{HP_{rev}} = COP_{HP_{rev}} + 1 \quad [3.6]$$

3.2. Receiving engines in the presence of internal irreversibilities

The engine represented by the 1-2-3-4 cycle (Figure 3.1) can be qualified as exoreversible: it has no heat transfer irreversibilities at the hot and cold ends.

It only has irreversibilities internal to the cycle:

– irreversible 4-1 and 2-3 isotherms;

– 1-2 and 3-4 adiabats (and not isentrops).

The literature then proposes two main approaches to describe the internal irreversibilities:

a) The *entropy analysis* method conforming to relation [3.2], which can also be written as:

$$\Delta S_{HS} = \Delta S_{CS} + \Delta SI \quad [3.7]$$

This method will be used in the rest of the development.

b) The *ratio* method introduced by Ibrahim (Ibrahim et al. 1991), among others, which can be written as:

$$\Delta S_{HS} = I_{CS}\, \Delta S_{CS} \quad [3.8]$$

Note that in this case, ΔS_{CS} is the entropy reference ($I_{CS} \geq 1$). In the general case, I_{CS} is a complex function to detail.

The influence of irreversibilities on the efficiency of the engine is obtained by combining relations [3.1] and [3.2], with the definition of efficiency. Only the case of the refrigerator is described below. The reader can find the results relating to the heat pump in the literature (Feidt 2016a). It becomes:

$$\frac{Q_{CS}+W}{T_{HS}} = \frac{Q_{CS}}{T_{CS}} + \Delta SI \qquad [3.9]$$

The COP of the refrigerator (relation [3.4]) therefore depends not only on the temperatures T_{HS}, T_{CS}, but also on the parameters (either Q_{CS} or W) and ΔSI.

To conclude:

$$COP_{RM} = COP_{RM_{rev}}\left[1 - \frac{T_{HS}\,\Delta SI}{W_0}\right] \qquad [3.10a]$$

or

$$\frac{1}{COP_{RM}} = \frac{1}{COP_{RM_{rev}}} + \frac{T_{HS}\,\Delta SI}{Q_{C0}} \qquad [3.10b]$$

The COP therefore depends on an additional constraint, by the parameters W_0 or Q_{C0} according to the initial observation by C.-H. Blanchard (1980), developed by Feidt (2010).

Relations [3.10a] and [3.10b] prove that the reversibility corresponds to the optimal conditions leading to the maximum COP. Other criteria will be introduced below. Note that at given W_0 or Q_{C0}, minimizing the internal entropy generation provides maximum COP in finite energy (W_0 or Q_{C0}). But reversibility presupposes quasi-staticity (cycle transformation time, and consequently cycle time tending towards infinity). The cooling capacity of the engine therefore tends towards zero.

Hence, a new approach called finite-time thermodynamics (FTT) is considered. Similar conclusions hold for the heat pump.

3.3. The Carnot refrigerator according to finite-time thermodynamics

C.H. Blanchard's (1980) article is one of the rare articles that consider transformation times and energies. So, the optimal COP_{HP} provided by the author conforms to:

$$\frac{1}{COP_{HP}} = 1 - \frac{T_{CS}}{T_{HS} + \left[1 + \sqrt{\frac{K_H}{K_C}}\right] \cdot \gamma \frac{\dot{Q}_H}{K_H}} \qquad [3.11]$$

with Δt being the cycle time such that:

$$\Delta t = \gamma(\Delta t_H + \Delta t_C)$$

where γ is a coefficient of proportionality of the isentropic duration to the hot Δt_H and cold Δt_C isotherm durations.

Note that in this description, the average heat rate exchanged at the hot end $\dot{Q}_H$ is dimensioned with respect to the average heat transfer conductance at the hot end (but averaged over the cycle).

This finding is fundamental, because unlike the article on the Carnot engine, there are two major consequences:

– For an endoreversible reverse cycle engine, the existence of a minimum mechanical power consumption is correlated with a heat rate constraint imposed on the cold end $\dot{Q}_C$ (RM-refrigerator) or the hot end $\dot{Q}_H$ (HP-heat pump).

Here, there is then a hidden assumption that the operation is in steady state (perhaps "finite speed", for Blanchard), but it is not explicit.

– The second consequence follows the previous one. The minimum energy consumption also corresponds to the maximum COP, given the additional constraint imposed.

This section concludes by stating that in the presence of an additional constraint, relations [3.10a] and [3.10b] prove that the maximum COP of a reverse cycle engine can be obtained with a maximum useful effect or energy consumption constraint under reversibility conditions (Gouy–Stodola theorem relating to reverse cycle engines).

It is up to the reader to show that, just as an exercise for a heat pump, the relationship between the COP_{actual} and the COP_{endo} in steady state conforms to:

$$COP_{HP} = COP_{HP_{rev}}\left[1 - \frac{T_{CS}\,\dot{S}_I}{\dot{W}_0}\right] \quad [3.12]$$

with $\dot{W}_0$ being the parameter (mechanical power):

$$\frac{1}{COP_{HP}} = \frac{1}{COP_{HP}} + \frac{T_{CS}\ \dot{S}_I}{\dot{Q}_{H0}} \quad [3.13]$$

with $\dot{Q}_{H0}$ being the parameter (useful heating value).

$\dot{S}_I$ represents the entropy generation rate of the heat pump.

Other efficacy criteria different from the COP have also been proposed (Wijeysundera 1999, Chapter 19; Feidt 2010; Petrescu and Costea 2011; Pop et al. 2012; Feidt 2013a). They will not be considered in this chapter, but may be revisited in *Efficiency in Practice*, the fourth book in this series. Among the other approaches that have also been proposed, the FST approach developed by Petrescu and Costea (2011) should be noted. Contrary to the article by C.H. Blanchard (1980), it is truly a finite speed thermodynamics: the proposed model shows a characteristic average speed as a variable of primordial influence on the irreversibility of reciprocating engines for mechanical vapor compression.

The following section proposes a general approach, while considering either a cycle energy approach or an energy rate approach associated with the stationary dynamic regime of the reverse cycle engine. The illustration is made on the reverse Carnot engine or its extensions.

3.4. The reverse cycle Carnot engine model according to finite physical dimensions thermodynamics (FPDT)

This model reflects an evolution of the work of the period (1990–2000) aimed at generalizing the models and unifying the descriptions (Feidt 2010).

3.4.1. *Model of a Carnot engine with thermal conductances*

The model is presented according to hypothesis 1 of the nominal steady state. The heat rate at the hot and cold thermostats is written as:

$$\dot{Q} = K.f(T_S, T) \quad [3.14]$$

K, the generalized heat transfer conductance;

T_S, the thermostat constant temperature;

T, the hot or cold isotherm temperature.

The literature then shows the following main heat transfer laws (algebraic convention):

– linear law: $f(T_S, T) = (T_S - T)$;

– radiative type law: $f(T_s, T) = T_s^n - T^n$ (order n);

– convective type law: $f(T_s, T) = (-1)^{n+1}(T_s - T)^n$;

– phenomenological type law: $f(T_s, T) = \left(\frac{1}{T} - \frac{1}{T_S}\right)$.

This chapter is limited to the linear law. For more details, the reader can refer to the literature (Feidt 1999).

The energy balance is therefore written as:

$$\dot{W} = \dot{Q}_H - \dot{Q}_C \qquad [3.15a]$$

Knowing that $\dot{Q}_H = \dot{Q}_{HS}$, $\dot{Q}_C = \dot{Q}_{CS}$ for a system without heat losses or gains (hypothesis 2), it becomes:

$$\dot{Q}_{HS} = K_H(T_H - T_{HS}); \;\; \dot{Q}_{CS} = K_C(T_{CS} - T_C) \qquad [3.15b]$$

The sign convention is as shown in Figure 3.2 (absolute values).

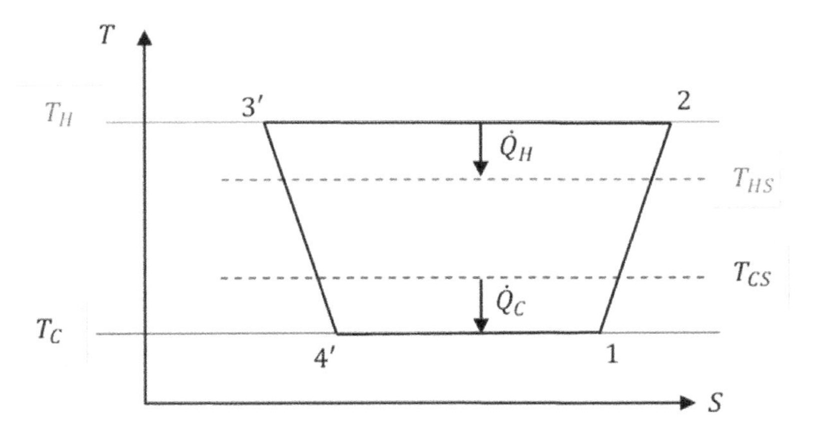

Figure 3.2. *Real reverse Carnot cycle*

The second law of thermodynamics leads to the following entropy balance for the cycled fluid:

$$\frac{\dot{Q}_C}{T_C} + \dot{S}_I = \frac{\dot{Q}_H}{T_H} \qquad [3.16]$$

In this model, T_H, T_C appear as unbounded variables (upper for T_H; lower for T_C by 0, which is unreachable). The entropy balance method is chosen according to relation [3.16] and not the ratio: the corresponding approach is more convenient to develop the entropy analysis or its extensions.

Finally, note that this thermal conductance model leads directly to the sizing of the heat transfer surfaces of both hot (A_H) and cold (A_C) exchangers, noting that:

$$K = k.A \qquad [3.17]$$

A, the heat transfer area;

k, the overall heat transfer coefficient between the thermostat and the cycled fluid. It is estimated by the heat transfer correlations available in the literature.

It should be noted that $K(k)$ are average cycle data, i.e. the most common in practice, which is not the case with the type of modeling recommended by Curzon and Ahlborn. In the latter case, the heat transfer conductance K' is a heat transfer conductance of the transformation of the cycle (in this case isothermal) such that

$$k'.\Delta t_T = K.\Delta t \qquad [3.18]$$

with Δt_H at the hot end and Δt_C at the cold end.

By taking the general form of relation [3.17], which characterizes the thermal conductance model, with the coefficient k (assumed to be constant), the system to be optimized appears as a system with finite geometric dimension (surface A).

If A_T is the total area available due to the size or cost, the problem to be solved is then an optimal distribution of A_T between the two hot and cold exchangers of the system. This problem is a generic problem. All that remains is to specify the chosen optimization criterion, coupled with the imposed geometric dimension constraint:

$$A_H + A_C = A_T \qquad [3.19]$$

Numerous results using this methodology are available in the literature. For an illustration of these results, please refer to Feidt (2014, pp. 567–572).

For a Carnot refrigerator, with the imposed useful effect $\dot{Q}_0$ and the finite dimensional constraint on the conductance (parameters K_T such that $K_H + K_C = K_T$):

$$k_{copt} = K_T \frac{1+\frac{\dot{S}_I}{K_T}}{2+T_{CS}\frac{\dot{S}_I}{\dot{Q}_0}} \qquad [3.20]$$

This expression shows that for a real refrigerator ($\dot{S}_I > 0$), conductance equipartition is only possible if $\dot{Q}_0 = \frac{K_T T_{CS}}{2}$.

Most of the time $\dot{Q}_0 < \frac{K_T T_{CS}}{2}$. Consequently, the search for the minimum mechanical power expenditure imposes $K_C < K_H$.

It is possible to search for the expression of the COP associated with this minimum power expenditure (at the discretion of the reader as an exercise). Another option, very little developed in the literature, can be chosen.

EXAMPLE.– COP_{RM} optimum for an irreversible refrigerator (Feidt 2014, pp. 567–572).

In this case, for both the temperature variables T_H, T_C:

$$T_{C_{opt}} = \frac{T_{CS}}{1+\sqrt{\frac{\dot{S}_I}{K_T}}} \quad [3.21]$$

$$T_{H_{opt}} = \frac{T_{HS}}{1-\sqrt{\frac{\dot{S}_I}{K_T}}} \quad [3.22]$$

Internal irreversibilities lead to temperature differences ($T_{H_{opt}} - T_{C_{opt}}$) that increase as the internal irreversibilities increase.

The maximum COP_{CE} leads to an optimal distribution of different transfer conductances:

$$MAX\ (COP_{RM}) = \frac{T_{CS}\left(1-\sqrt{\frac{\dot{S}_I}{K_T}}\right)^2}{T_{HS}\left(1+\sqrt{\frac{\dot{S}_I}{K_T}}\right)^2 - T_{CS}\left(1-\sqrt{\frac{\dot{S}_I}{K_T}}\right)^2} \quad [3.23]$$

The trends in this example are the same as in the previous example; however, in the second example the optimal conductance distribution only depends on $\dot{S}_I$ and K_T.

To conclude this section, readers that aim to become engineers are advised to consult a study on the development of this approach, integrating the specificities of refrigerators with mechanical vapor compression (presence of desuperheating before the condensation phase; isenthalpic expansion) (Pop et al. 2012). The reported results show the evolution of the COP for various cold powers according to the transfer conductance, and this for different refrigerants.

3.4.2. *Immediate extensions of the model with thermal conductances*

3.4.2.1. *Other criteria*

As previously mentioned, the COP criterion corresponds to the natural criterion of the first law of thermodynamics. This criterion is interesting because it is dimensionless.

But there are others. Among these are mentioned:

– exergy efficiency;

– efficiency in the sense of the second law or quality factor.

Quite recently, the so-called ECOP thermo-ecological criterion has appeared (Ust and Sahin 2017). Then, multi-criteria analyses.

Despite the importance of the economy and the environment, this discourse is limited to the energy objective functions in accordance with the diagram in Figure 3.3.

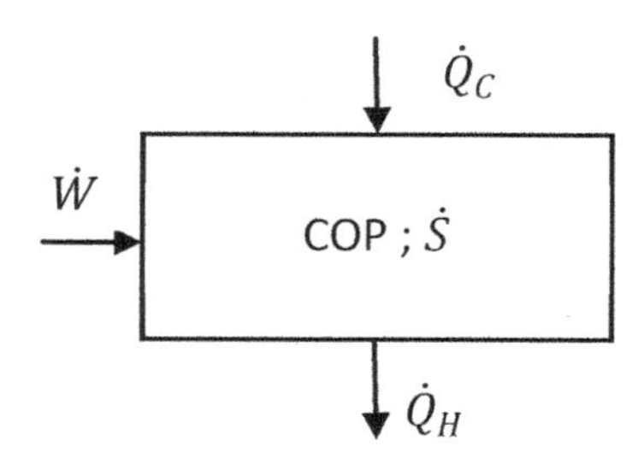

Figure 3.3. *Block diagram of a reverse cycle engine*

The major objective functions that appear in Figure 3.3 are:

– maximum useful effect ($\dot{Q}_C$ for a refrigerator; $\dot{Q}_H$ for a heat pump);

– minimum energy consumption, $\dot{W}$;

– maximum efficiency in the sense of the first law of thermodynamics (COP);

– minimum heat rejection into the environment for a refrigerator, or maximum heat removed from the environment for a heat pump;

– minimum entropy generation rate for the reverse cycle engine ($\dot{S}_I$) or for the system (engine + source + sink) $\dot{S}_S$.

3.4.2.2. *Optimization, constrained or not*

The first optimization of reverse cycle engines is based on the energy balance equation [3.1] and the entropy balance equation [3.2]. The natural variables that then appear are the intensive quantities, namely temperature T_H, T_C, unbounded, unlike engines.

The problem therefore has temperatures as degrees of freedom. But the additional constraint induces finiteness in temperature. Thus, the additional constraint is most often an extensive constraint, either $\dot{W}_0$ (MEE imposed) or $\dot{Q}_0$ (UCE imposed), eventually imposed COP_0 (imposed efficiency). The corresponding results are summarized in Petrescu et al. (2015, pp. 435–440).

3.4.2.3. *Influence of entropy generation*

The results shown so far were relative to the case where the entropy generation rate of the cycle was a constant parameter. Assume entropy generation functions as follows:

– $\dot{S}_I = s_I(T_H - T_C)$

– $\dot{S}_I = c_I \ln\left(\frac{T_H}{T_C}\right)$

It follows that the influence of the form of the entropy generation function of the reverse cycle engine does not modify the main trends observed on the optimization results. Additional investigations would nevertheless be desirable so as to confirm the general nature of these conclusions.

Before ending this section, it should be specified that as it stands, the entropy analysis leads to a minimization of the entropy generation of the system (engine + thermostats) $\dot{S}_S$ such that:

$$\dot{S}_S = \dot{S}_I + K_H \frac{(T_H - T_{HS})^2}{T_H T_{HS}} + \frac{K_C (T_{CS} - T_C)^2}{T_C T_{CS}} \qquad [3.24]$$

It is then easy to see that the minimum of $\dot{S}_S$ differs from that of $\dot{W}$, or from the minimum of UCE whatever the form of function $\dot{S}_I(T_H, T_C)$, which limits the interest of minimizing entropy generation.

3.5. Generalization of the reverse cycle Carnot engine model according to FPDT

Relation [3.14] was the introductory relationship to the thermal conductance heat transfer model associated with the thermostatic hot and cold reservoirs. This case is well suited to the phase change heat transfer of a pure substance. However, in the case where the heat transfers at the hot and cold source involve sensible heat, the transfer model is to be resumed in the following form:

$$\dot{Q} = \varepsilon\, \dot{c}_{min}(T_{Se} - T) \qquad [3.25]$$

with T_{Se}, the finite heat exchanger inlet temperature;

T_{SS}, the finite temperature at the outlet of the exchanger such that:

$$\dot{Q} = \dot{C}_{min}|T_{Se} - T_{SS}| \qquad [3.26]$$

This last relation allows the determination of T_{SS}.

The particularization of relation [3.26] to the reverse Carnot cycle induces:

$$\dot{Q}_H = \varepsilon_H\, \dot{c}_H(T_H - T_{HSe});\ \dot{Q}_C = \varepsilon_C\, \dot{c}_C(T_{CSe} - T_C) \qquad [3.27]$$

with $\dot{c}_H = \dot{m}_H c_{PH}$, the heat capacity rate of the hot source fluid;

$\dot{c}_C = \dot{m}_C c_{PC}$, the heat capacity rate of the cold source fluid;

(c_{Pi}, specific heat of fluid i; $\dot{m}_i$, its mass flow rate).

A comparison of relationships [3.15b] and [3.27] allows the generalization of the model, with respect to a finite physical variable G_i such that:

$$\dot{Q}_H = G_H(T_H - T_{HSe});\ \dot{Q}_C = G_C(T_{CSe} - T_C) \qquad [3.28]$$

This generalization of the linear case introduces finite physical variables, generalized heat transfer conductance (W/K) such as:

$$G_H + G_C = G_T \qquad [3.29]$$

It is then possible to use the previous methodology to perform a simultaneous (or sequential) optimization using the variational tool. Depending on the complexity of the problem, the need for digitization using the computer tool arises early on.

Works on this topic have already been published (Feidt and Haberschill 2019; Feidt and Costea 2020) and others are still in development.

It should be noted that the physical variable G_i introduced corresponds to the heat transfer conductances initially introduced in section 3.4, as well as to the variable $\varepsilon\dot{c}_{min}$ (this section 3.5).

It is then possible, with an imposed finite source and sink ($\dot{c}_H, \dot{c}_C$ parameters), to find the optimal distribution of exchanger efficiencies, with new finite variables subject to the constraint:

$$\varepsilon_H + \varepsilon_T = \varepsilon_T \leq 2 \qquad [3.30]$$

Also note that for the reverse Carnot cycle engine, the efficiency ε that hot and cold exchangers satisfy is:

$$\varepsilon = 1 - \exp(-NUT)$$

with $NUT = \frac{K}{\dot{c}_{min}}$.

With $\dot{c}_{min}$ being the optimization parameter, the previous equations can lead to the optimization of the transfer conductance distribution, but this time for finite hot and cold sources.

Still, with $\dot{c}_{min}$ as the parameter, the general expression of the conductance K given by relation [3.17] always leads to the optimal distribution of the transfer surfaces under the constraint [3.19], where the overall heat transfer coefficients k are assumed to be constant parameters.

A final step would involve seeking the optimal distribution of the heat capacity rates (or mass flow rates) of the external fluids circulating at the hot end and cold end of the engine with the constraint:

$$\dot{c}_H + \dot{c}_C = \dot{c}_T$$

or

$$\dot{m}_H + \dot{m}_C = \dot{m}_T$$

Work on this topic is in progress.

3.6. Latest advances in a reverse cycle Carnot engine model

3.6.1. *Energy model*

Very recently, a renewed approach to modeling inverse cycle engines was proposed, based on the Carnot cycle, but explicitly introducing a coupling constraint between the reservoirs and the engine (Novikov 1957; Feidt 2013b, pp. 334–342). Here, the main ideas and results relating to a refrigerator are summarized, with the focus on the conjugate variables of the cycle (Figure 3.4). We explore the energies exchanged W and Q.

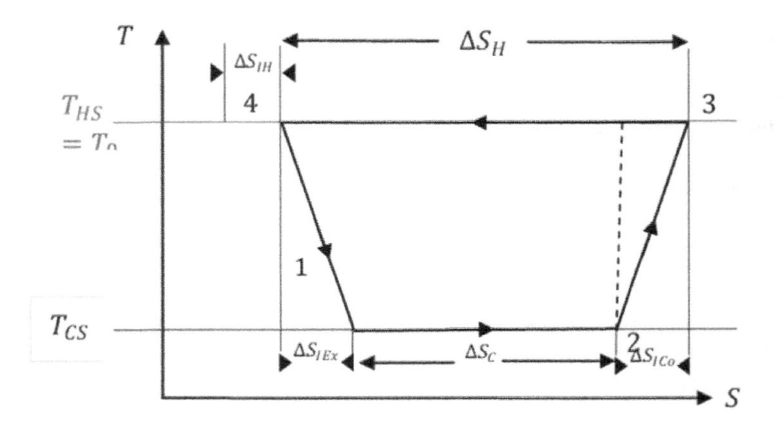

Figure 3.4. *Carnot endo-irreversible refrigerator*

Thus, corresponding to [3.15] and without heat gains or losses:

$$W = Q_H - Q_C \quad [3.31]$$

with $Q_H = T_{HS}\,\Delta S_H$ ($T_{HS} = T_0$ for the refrigerator)

$$Q_C = T_{CS}\,\Delta S_C$$

The entropy balance remains analogous to relation [3.16]:

$$\Delta S_H = \Delta S_C + \Delta S_I \quad [3.32]$$

The entropy reference for the refrigerator is the heat transfer entropy at the cold end ($\Delta S_C = \Delta S$). Combining equations [3.31] and [3.32] yields:

$$W = \Delta S(T_{HS} - T_{CS}) + T_0 \Delta S_I \quad [3.33]$$

This relationship corresponds to the Gouy–Stodola theorem applied to refrigerators: the minimum energy consumption corresponds to the minimum entropy generation in an endoreversible engine.

Also note that the presence of useful transfer entropy at the cold source is a condition for the existence of the cycle, and therefore of the system. It is a constraint of the problem such that: $W_{rev} = \Delta S(T_{HS} - T_{CS})$.

The use of relations [3.4], [3.32] and [3.33] provides the expression of COP_{RM} in a complementary form to relations [3.10a] and [3.10b]:

$$\frac{1}{COP_{RM}} = \frac{1}{COP_{RMrev}} + \frac{T_0}{T_{CS}} \frac{\Delta S_I}{\Delta S} \quad [3.34]$$

The COP of the real refrigerator decreases by two ratios:

– the intensive ratio $\frac{T_0}{T_{CS}}$;

– the extensive ratio $\frac{\Delta S_I}{\Delta S}$, which corresponds to the ratio introduced by I. Novikov for Carnot engines (Novikov 1957).

3.6.2. *Minimizing the energy expenditure of the Carnot refrigerator (power)*

Now moving onto FTT, the average power absorbed over a cycle is written as:

$$\overline{\dot{W}} = \frac{W_{rev} + T_0 \Delta S_I}{\Delta t} \quad [3.35]$$

Assuming that the entropy generation over the cycle is inversely proportional to Δt, then:

$$\Delta S_I = \frac{C_I}{\Delta t} \quad [3.36]$$

Equation [3.36] brings out an important concept: C_I (JsK^{-1}) is an entropy-generating action, by analogy to the action defined in classical mechanics (Js). $\Delta t \to \infty$ corresponds to the reversibility of a quasi-static cycle.

The consideration of relation [3.35] shows that at a given W_{rev}, the average power consumed is a decreasing function of the cycle time. On the contrary, if the average cooling power of the refrigerator is fixed, the energy consumption is provided by the modified form of the following relation [3.35]:

$$W = \overline{\dot{W}}_{rev}\, \Delta t + \frac{C_I T_0}{\Delta t} \quad [3.37]$$

with $\overline{\dot{W}}_{rev} = (T_{HS} - T_{CS})\dot{S}\Delta t$.

In this case, the minimum MEE occurs for a cycle duration Δt^* such that:

$$\Delta t^* = \sqrt{\frac{C_I T_0}{\overline{\dot{W}_{rev}}}} \quad [3.38]$$

The minimum equals:

$$\min(W) = 2\sqrt{\overline{\dot{W}}_{rev} C_I T_0} \quad [3.39]$$

3.6.3. *The modified Chambadal refrigerator*

Figure 3.5 shows the modified Chambadal refrigerator, since it includes internal irreversibilities in addition to the irreversibility of the heat transfer at the source between T_{CS} and T_C.

3.6.3.1. *Model without a source coupling equation*

In this case (Feidt and Costea 2020):

$$\frac{1}{COP_{CE}} = \frac{T_0}{T_{CS}}.F - 1 \quad [3.40]$$

with F being the corrective factor with respect to the reversible situation such that:

$$F = \frac{T_{CS}}{T_C}\left(1 + \frac{\Delta S_I}{\Delta S}\right) \quad [3.41]$$

Note that the corrective factor depends on an intensity ratio (irreversibility due to the thermal gradient of heat transfers at the source) and an extensity ratio accounting for the importance of internal irreversibilities in the refrigerator.

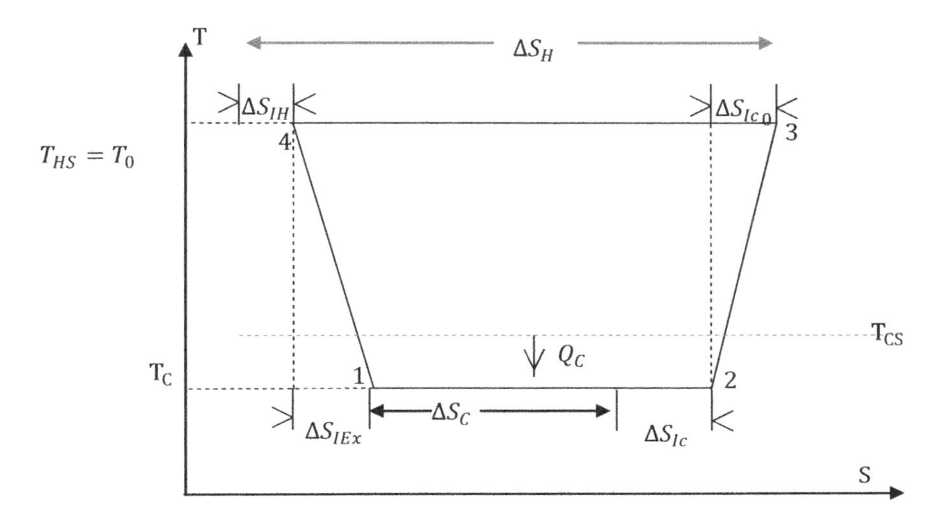

Figure 3.5. *Representation of the modified Chambadal refrigeration cycle*

3.6.3.2. *Model with a source coupling equation*

With the assumption of a linear heat transfer law, it comes for the cold source, in accordance with relation [3.28]:

$$Q_C = T_C \Delta S = G_C (T_{CS} - T_C) \quad [3.42]$$

This new constraint [3.42] makes T_C a dependent variable according to:

$$T_C = T_{CS} \frac{G_C}{G_C + \Delta S} \quad [3.43]$$

Using the same approach as in section 3.6.3.1, F is expressed as follows:

$$F = \frac{G_C + \Delta S}{G_C}\left(1 + \frac{\Delta S_I}{\Delta S}\right) \quad [3.44]$$

In the endoreversible case, it becomes:

$$\frac{1}{COP_{CEendo}} = \frac{T_0}{T_{CS}} \cdot \frac{G_C + \Delta S}{G_C} - 1 \quad [3.45]$$

Equation [3.45] constitutes a new upper bound to the COP for an endoreversible Chambadal refrigerator: the ratio $\frac{\Delta S}{G_C}$ decreases the COP when it increases.

3.6.4. *The modified Curzon–Ahlborn refrigerator*

Figure 3.6 shows the modified Curzon–Ahlborn refrigerator. This refrigerator includes both internal irreversibilities and heat transfer irreversibilities at the cold source and heat sink (between T_H and T_{HS}).

The model without a coupling constraint was reported in section 3.3. Therefore, only the extensions to section 3.1 will be discussed here.

The coupling constraint at the source remains and therefore so does equation [3.43], but the same type of formal constraint appears at the hot end:

$$Q_H = T_H \Delta S_H = G_H(T_H - T_{HS}) \qquad [3.46]$$

Consequently, the use of equations [3.46] and [3.7] leads to the following expression for the dependent variable T_H:

$$T_H = T_{HS} \frac{G_H}{G_H - (\Delta S + \Delta S_I)} \qquad [3.47]$$

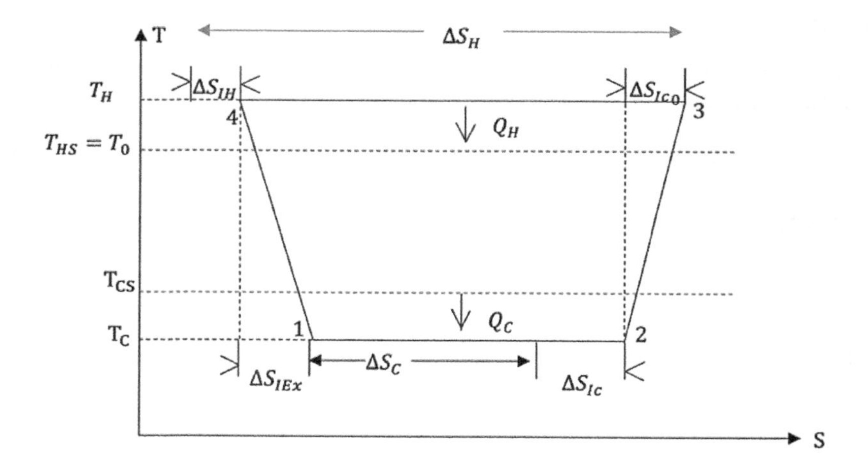

Figure 3.6. *Representation of the modified Curzon–Ahlborn refrigeration cycle*

Assuming ΔS_I to be a constant parameter, and the constraint of finite physical dimension (relation [3.29]), the variational calculation leads to the optimal distribution of the thermal energy transfer conductance in analytical form:

$$G_C^* = \frac{\Delta S}{(\Delta S+\Delta S_I)\sqrt{T_{HS}}+\Delta S\sqrt{T_{CS}}}\left((G_T - \Delta S - \Delta S_I)\sqrt{T_{CS}} - (\Delta S + \Delta S_I)\sqrt{T_{HS}}\right) \quad [3.48]$$

$$G_H^* = \frac{\Delta S+\Delta S_I}{(\Delta S+\Delta S_I)\sqrt{T_{HS}}+\Delta S\sqrt{T_{CS}}}\left((G_T + \Delta S)\sqrt{T_{HS}} + \Delta S\sqrt{T_{CS}}\right) \quad [3.49]$$

The minimum energy consumption is deduced as follows:

$$\min W = \frac{1}{G_T-\Delta S_I}\left(G_T(T_{HS}(\Delta S + \Delta S_I) - T_{CS}\Delta S) + \Delta S(\Delta S + \Delta S_I)\left(\sqrt{T_{HS}} + \sqrt{T_{CS}}\right)^2\right) \quad [3.50]$$

This function is an increasing function of ΔS and ΔS_I, but a decreasing function of G_T, which is related to the investment cost and the size of the system. There is therefore an economic compromise between the investment cost and the operating cost.

To complete the results, the COP_{MAF} corresponding to $\min(W)$ becomes:

$$\frac{1}{COP_{RM}(\min W)} = \frac{T_{HS}}{T_{CS}}.F - 1 \quad [3.51]$$

with

$$F = \frac{\Delta S+\Delta S_I}{\Delta S}\frac{G_T\sqrt{T_{HS}}+\Delta S\left(\sqrt{T_{HS}}+\sqrt{T_{CS}}\right)}{G_T\sqrt{T_{CS}}-(\Delta S+\Delta S_I)\left(\sqrt{T_{HS}}+\sqrt{T_{CS}}\right)}\sqrt{\frac{T_{CS}}{T_{HS}}} \quad [3.52]$$

The corrective factor therefore depends not only on the extensities via the ratios $\frac{\Delta S_I}{\Delta S}$ and $\frac{\Delta S}{G_T}$, but also on intensities via the ratio $\sqrt{\frac{T_{CS}}{T_{HS}}}$. The limit COP of equilibrium thermodynamics is found when both $\frac{\Delta S_I}{\Delta S}$ and $\frac{\Delta S}{G_T}$ tend towards zero. The ratio $d_I = \frac{\Delta S_I}{\Delta S}$ that appears here will be used in the following. It consists of a ratio of entropic quantities, namely between the overall entropy generation of the cycle and its reference entropy ΔS (see equations [3.32] and [3.33]).

If $\frac{\Delta S_I}{\Delta S} = 0$, expression [3.51] provides a new endoreversible limit for the COP_{RM} at $\min(W)$. A similar expression exists for the heat pump.

Another fundamental consequence relates to the total entropy generation of the system ΔS_{I_S} provided by an expression similar to relation [3.24]:

$$\Delta S_{I_S} = \Delta S\left[\frac{\Delta S}{G_C+\Delta S} + d_I + \frac{\Delta S(1+d_I)}{G_H-\Delta S(1+d_I)}\right] \quad [3.53]$$

This function increases by ΔS and d_I, at fixed G_H and G_C. But it also presents an extremum for:

$$G_C^* = \frac{G_T - 2\Delta S(1+d_I)}{2+d_I} \quad [3.54]$$

$$G_H^* = \frac{G_T + 2\Delta S}{2+d_I} \quad [3.55]$$

The corresponding optimum is:

$$Opt(\Delta S_{I_S}) = \Delta S\left[d_I + \frac{\Delta S(2+d_I)^2}{G_T - d_I \Delta S}\right] \quad [3.56]$$

But above all, the comparison of relations ([3.48], [3.49]) and ([3.54], [3.55]) clearly shows that the optimal distributions of thermal energy transfer conductance differ. The method of minimizing the entropy generation is therefore not suitable for reverse cycle engines.

In the endoreversible case ($d_I = 0$), the entropy generation is a quadratic function of ΔS and a decreasing function of G_T. The equilibrium thermodynamic limit occurs if $\frac{\Delta S}{G_T}$ tends towards zero.

3.7 Extension of finite physical dimensions thermodynamics to two complex systems

This is to show that the detailed method for simple two-reservoir engines extends to more complex configurations.

3.7.1. *Complex two-reservoir systems*

As an example, the point in section 3.7.1 will be discussed using an optimization of an endoreversible two-stage heat pump without heat gains or losses (section 3.7.1.1), then a double-function heat pump will be considered (section 3.7.1.2).

3.7.1.1. *Optimization of a two-stage heat pump*

Figure 3.7 represents the two interlocked cycles of an endoreversible engine without heat gains or losses. The model is a conductance model $K_i(WK^{-1})$ at nominal conditions, using a linear transfer law.

The objective function is the mechanical power $\dot{W}$ for which the minimum is sought in the presence of the heating effect constraint imposed $\dot{Q}_H = \dot{Q}_0$, which amounts to the same as the search for the maximum of COP_{HP}:

$$COP_{HP} = \frac{K_H(T_H - T_{HS})}{K_H(T_H - T_{HS}) - K_C(T_{CS} - T_C)} \quad [3.57]$$

The reader will verify that the temperature optimization includes the variables (T_H, T'_H, T_C, T'_C), five parameters $(T_{HS}, T_{CS}, K_H, K', K_C)$ and the objective function (relation [3.57]).

In fact, the objective function is expressed in terms of the variables T_H, T_C and the ratio of intermediate temperatures. The entropy balance constraints and the useful effect constraint provide the three previous variables depending on the parameters of the problem.

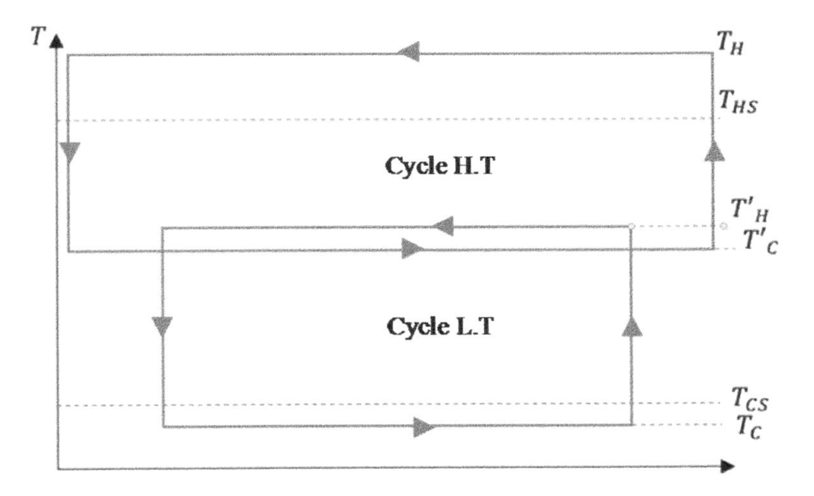

Figure 3.7. *LT and HT cycles of a two-stage endoreversible engine*

The optimization of the thermal conductance distribution leads to the following objective function:

$$\dot{W} = \dot{Q}_0 \left[1 - \frac{T_{CS}}{T_{HS} + \dot{Q}_0 \left(\frac{1}{K_C} + \frac{1}{K'} + \frac{1}{K_H} \right)} \right] \quad [3.58]$$

with the finiteness constraint:

$$K_C + K' + K_H = K_T \qquad [3.59]$$

Here, endoreversibility leads to the equipartition of K_i such that:

$$MAX\left(COP_{HP_{endo}}\right) = \frac{1}{1 - \frac{T_{CS}}{T_{HS} + 9\frac{\dot{Q}_0}{K_T}}} \qquad [3.60]$$

These results are currently being extended.

3.7.1.2. *Optimization of a double-function heat pump*

This optimization concerns the simultaneous use of heat and cold in the food industry, using reverse cycle engines with mechanical vapor compression.

The proposed model is again dedicated to the steady state regime without heat gains or losses to the atmosphere T_0 (Figure 3.8), but in the presence of internal irreversibilities in the engine $\dot{S}_I$ (finite conductance model K_i).

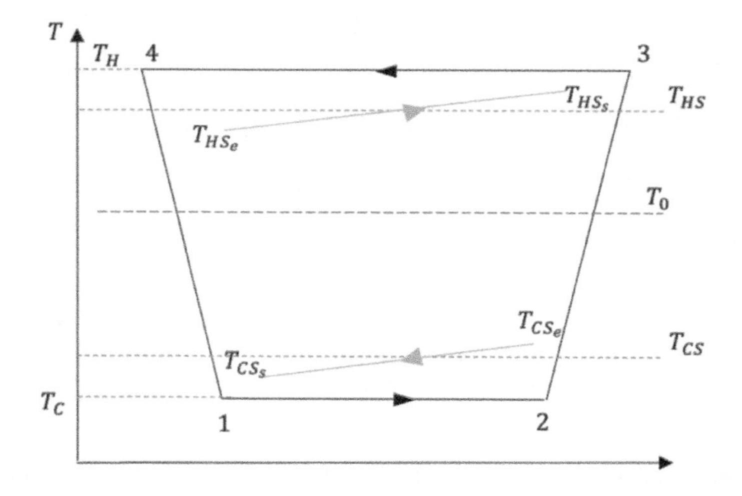

Figure 3.8. *Double-function heat pump model with internal irreversibility*

It is important to note that the ambient temperature T_0 ranges between the temperatures at the hot end (upper) and at the cold end (lower).

The equations of the model, now well known, are not repeated. The thermokinetic approach proceeds according to the linear transfer law, but by introducing X_H, X_C such that:

$$\dot{Q}_H = K_H(T_H - T_{HS}) = K_H X_H \qquad [3.61]$$

$$\dot{Q}_C = K_C(T_{CS} - T_C) = K_C X_C \qquad [3.62]$$

X_H, X_C should not be confused with the pinch at the condenser $(T_H - T_{HS_S})$ and the evaporator $(T_{CS_S} - T_C)$. The finite physical dimension is relative to the thermal conductances for which the optimal distribution is sought.

Two efficiency criteria according to thermodynamics appear:

1) The COP of the double-function heat pump according to the first law of thermodynamics

Recall the two definitions relating to the refrigerator and the heat pump:

$$COP_{IRM} = \frac{\dot{Q}_C}{\dot{W}}; \; COP_{IHP} = \frac{\dot{Q}_H}{\dot{W}}$$

The use of the first law then leads to the useful effect of the double-function heat pump (DHP) for the same energy consumption:

$$EU_{DHP} = \dot{Q}_C + \dot{Q}_H$$

Hence, the COP_{IDHP}:

$$COP_{IDHP} = \frac{\dot{Q}_H + \dot{Q}_C}{\dot{Q}_H - \dot{Q}_C} = COP_{IRM} + COP_{IHP} \qquad [3.63]$$

An important remark is that the two useful effects occur at different temperature levels, which deserves consideration.

2) The COP of the double-function heat pump according to the second law

It is possible to define several coefficients, depending on the references (see *Efficiency in Practice* for more details). Here, the definition of the COP_{ITFP} is limited to exergy analysis:

$$COP_{Ex_{DHP}} = -\frac{\dot{E}x_H + \dot{E}x_C}{\dot{W}} \qquad [3.64]$$

with

$$\dot{E}x_H = \dot{Q}_H\left(1 - \frac{T_0}{T_{HS}}\right); \dot{E}x_C = \dot{Q}_C\left(1 - \frac{T_0}{T_{CS}}\right).$$

The criterion makes it possible to take into account both the quality (intensity) and the quantity of energy.

The use of variational calculus leads to:

$$COP_{IDHP} = \frac{\left(1+\sqrt{\frac{\dot{S}_I}{K_T}}\right)^2 T_{HS}+\left(1-\sqrt{\frac{\dot{S}_I}{K_T}}\right)^2 T_{CS}}{\left(1+\sqrt{\frac{\dot{S}_I}{K_T}}\right)^2 T_{HS}-\left(1-\sqrt{\frac{\dot{S}_I}{K_T}}\right)^2 T_{CS}} \quad [3.65]$$

with $K_H^* = \frac{1}{2}\left(K_T + \frac{\dot{S}_I}{\alpha^*}\right)$; $K_C^* = \frac{1}{2}\left(K_T - \frac{\dot{S}_I}{\alpha^*}\right)$; $X_H^* = \frac{\alpha^*}{1-\alpha^*}T_{HS}$; $X_C^* = \frac{\alpha^*}{1+\alpha^*}T_{CS}$ and $\alpha^* = \sqrt{\frac{\dot{S}_I}{K_T}}$.

Likewise:

$$COP_{ExDHP} = \frac{(T_{HS}-T_0)\left(1+\sqrt{\frac{\dot{S}_I}{K_T}}\right)^2 -(T_{CS}-T_0)\left(1-\sqrt{\frac{\dot{S}_I}{K_T}}\right)^2}{T_{HS}\left(1+\sqrt{\frac{\dot{S}_I}{K_T}}\right)^2 -T_{CS}\left(1-\sqrt{\frac{\dot{S}_I}{K_T}}\right)^2} \quad [3.66]$$

In both cases, the quantities denoted above with a star (K_H^*, K_C^*, X_H^*, X_C^* and α^*) remain the same, but the energy and exergy efficiencies differ at the optimum.

Approaching the endoreversible case limit ($\dot{S}_I \to 0$), the thermostatic limits are:

$$\lim_{S_I \to 0} COP_{IDHP} = \frac{T_{HS}+T_{CS}}{T_{HS}-T_{CS}}; \ \lim_{S_I \to 0} COP_{ExDHP} = 1.$$

3.7.2. *Some comments on reverse cycle engines with three and four reservoirs*

Figure 3.9 is a basic diagram of a geothermal sorption heat pump (or a refrigerator or a sorption air conditioner). The same diagram would be valid for a solarized system, where the geothermal source is replaced by a solar one with various possible solar collectors according to the desired temperature level.

It can be seen from the diagram in Figure 3.9 that the engine has five heat exchangers, including a regenerative exchanger between a weak solution (low refrigerant concentration) but hot, and a strong solution (high refrigerant concentration) but cold.

In addition, the diagram shows a hot source (geothermal) at T_{HS}, a cold source at T_{CS} and then an ambient heat sink at intermediate temperature T_{IS}. A steady-state operation requires some mechanical energy $\dot{W}$ for the lift pump. Whatever the useful effect of the system, it appears that it is a three-heat-source engine, with here the hypothesis of thermostats (T_{HS}, T_{IS}, T_{CS}).

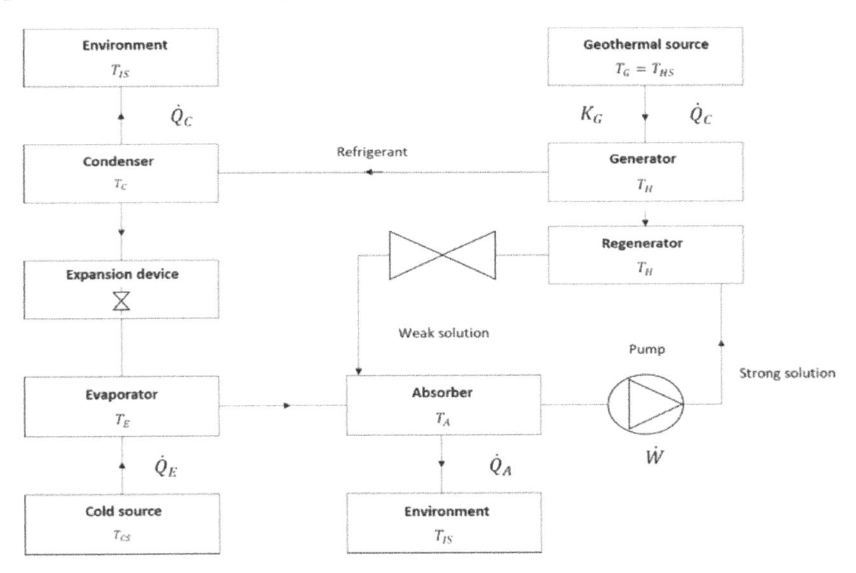

Figure 3.9. *Block diagram of a geothermal sorption engine*

The energy and entropy balances are written as:

$$\dot{Q}_C + \dot{Q}_A = \dot{Q}_G + \dot{Q}_E + \dot{W} \quad [3.67]$$

$$\frac{\dot{Q}_C + \dot{Q}_A}{T_{IS}} = \frac{\dot{Q}_G}{T_{HS}} + \frac{\dot{Q}_E}{T_{CS}} + \dot{S}_I \quad [3.68]$$

The definitions of COP RM and HP, for refrigerators and heat pumps respectively, are:

$$COP_{RM} = \frac{\dot{Q}_E}{\dot{Q}_G + \dot{W}} \quad [3.69]$$

$$COP_{HP} = \frac{\dot{Q}_I}{\dot{Q}_C+\dot{W}} \qquad [3.70]$$

with $\dot{Q}_I = \dot{Q}_C + \dot{Q}_A$ and $T_{IS} = T_0$ for RM, $T_{CS} = T_0$ for HP.

NOTE.– The case of heat transformers (see Feidt 1999) is not discussed here. The combination of equations [3.67]–[3.70] provides the COPs of real three-heat-source engines:

$$COP_{RM} = \frac{T_{CS}^{-1}-T_{HS}^{-1}}{T_{CS}^{-1}-T_{IS}^{-1}}\left[1+\frac{\dot{W}-T_{HS}\dot{S}_I}{\dot{W}+\dot{Q}_G}\cdot\frac{T_{CS}}{T_{HS}-T_{CS}}\right] \qquad [3.71]$$

$$COP_{HP} = \frac{T_{IS}^{-1}-T_{HS}^{-1}}{T_{CS}^{-1}-T_{IS}^{-1}}\left[1+\frac{\dot{W}-T_{HS}\dot{S}_I}{\dot{W}+\dot{Q}_G}\cdot\frac{T_{IS}}{T_{HS}-T_{IS}}\right] \qquad [3.72]$$

Note that relations [3.71] and [3.72] are expressed as a function of the heat input required $\dot{Q}_G$ (often the constraint imposed).

3.7.2.1. *Towards an optimal distribution of the heat transfer areas in the three-heat-source engine*

Going beyond the use of a linear heat transfer law, a general form for the three exchanged heat fluxes can be written as:

$$\dot{Q}_i = k_i A_i f(T_{is}, T_i) \qquad [3.73]$$

Equations [3.67] and [3.68] and the finiteness theorem in the transfer area ($\Sigma_i A_i = A_T$) allow the expression of the transfer surfaces according to the parameters ($T_{HS}, T_{IS}, T_{CS}; k_H, k_I, k_C; \dot{W}, \dot{S}_I$) and temperature variables (T_H, T_I, T_C) as follows:

$$A_H = \frac{1}{\Delta}\left[A_T k_I f_I k_C f_C\left(\frac{1}{T_C}-\frac{1}{T_I}\right)+\dot{W}\left(\frac{k_C f_C}{T_C}-\frac{k_I f_I}{T_I}\right)-\dot{S}_I(k_C f_C - k_I f_I)\right]$$

$$A_I = \frac{1}{\Delta}\left[A_T k_C f_C k_H f_H\left(\frac{1}{T_H}-\frac{1}{T_C}\right)+\dot{W}\left(\frac{k_H f_H}{T_H}-\frac{k_C f_C}{T_C}\right)-\dot{S}_I(k_H f_H - k_C f_C)\right] \qquad [3.74]$$

$$A_C = \frac{1}{\Delta}\left[A_T k_H f_H k_I f_I\left(\frac{1}{T_I}-\frac{1}{T_H}\right)+\dot{W}\left(\frac{k_I f_I}{T_I}-\frac{k_H f_H}{T_H}\right)-\dot{S}_I(k_I f_I - k_H f_H)\right]$$

with $\Delta = k_I f_I k_C f_C\left(\frac{1}{T_C}-\frac{1}{T_I}\right)+k_C f_C k_H f_H\left(\frac{1}{T_H}-\frac{1}{T_C}\right)+k_H f_H k_I f_I\left(\frac{1}{T_I}-\frac{1}{T_H}\right)$.

Relations [3.74] show the influence of $\dot{W}, \dot{S}_I$ on the optimal distribution of transfer areas. However, there remains margins of optimization in relation to variable temperatures: digitization is essential. If $\dot{W}$ is neglected, the endoreversible case leads to a much simpler, but numerical, approach. The results are available in the literature (Feidt 2013b, pp. 229–234; Feidt 2016b, pp. 398–404).

3.8. Some conclusions and perspectives

Using the proposed methodology, whatever the case studied, the main consequences are:

– highlighting the existence of optimal situations associated with compromises due to the existence of various physical dimensions including durations;

– the existence of new forms for the upper bounds of the efficiency of reverse cycle engines, due to various objectives or constraints (including boundary and initial conditions);

– the explanation of the links with the equipartition theorem, and also the shortcomings of the so-called method of minimizing the entropy generation;

– its extended application in engineering: future extensions towards economic and environmental optimization;

– design optimization is well advanced (nominal state). The optimization of operation (control command of transitions) remains to be further developed.

There are, however, already promising results (Stanciu et al. 2015, pp. 449–459).

3.9. References

Blanchard, C.H. (1980). Coefficient of performance for finite speed heat pump. *J. Appl. Phys.*, 51, 2471–2472.

Carnot, S. (1824). *Réflexion sur la puissance motrice du feu et sur les machines propres à développer cette puissance.* Bachelier, Paris.

Feidt, M. (1999). Thermodynamics and optimization of reverse cycle machines. In *Thermodynamics and Optimization of Complex Energy Systems*, Bejan, A. and Mamut, E. (eds). Kluwer Academic Press, Dordrecht.

Feidt, M. (2010). Thermodynamics applied to reverse cycle machine, a review. *Int. J. Refrig.*, 33(7), 1327–1342.

Feidt, M. (2013a). Evolution of thermodynamic modelling for three or four heat reservoirs reverse cycle machine, a review and new trends. *Int. J. Refrig.*, 36, 8–23.

Feidt, M. (2013b). *Thermodynamique optimale en dimensions physiques finies.* Hermes Lavoisier, Paris.

Feidt, M. (2014). *Génie énergétique. Du dimensionnement des composants au pilotage des systèmes*. Dunod, Paris.

Feidt, M. (2016a). From equilibrium thermodynamics to finite dimensions. Non-equilibrium thermodynamics of a heat pump. *COFRET'16*, 51–006, Bucharest.

Feidt, M. (2016b). *Thermodynamique et optimisation énergétique des systèmes et procédés*. TEC & DOC, Paris.

Feidt, M. and Costea, M. (2020). Effect of machine entropy production on the optimal performance of a refrigerator. *Entropy*, 22, 913.

Feidt, M. and Haberschill, P. (2019). Memo froid. *Rev. Gen. Froid Cond. Air.*, 117(3), 12–16.

Ibrahim, O.M., Klein, S.A., Mitchell, J.W. (1991). Optimum heat power cycles for specified boundary conditions. *J. Eng. Gas Turb. Power*, 113(4), 514–521.

Novikov, I.I. (1957). Efficiency of an atomic power generation installation. *Sov. J. Atom. Energy*, 3(11), 1269–1272.

Petrescu, S. and Costea, M. (2011). *Development of Thermodynamics with Finite Speed and Direct Method*. AGIR Publishing House, Bucharest.

Petrescu, S., Costea, M., Feidt, M., Ganea, I., Boriaru, N. (eds) (2015). *Advanced Thermodynamics of Irreversible Processes with Finite Speed and Finite Dimensions*. AGIR Publishing House, Bucharest.

Pop, H., Popescu, T., Popescu, G., Baran, N., Feidt, M., Apostol, V. (2012). Optimization model of single stage vapor compression refrigeration systems. *U.P.B. Sci. Bull. D*, 74(3), 91–106.

Stanciu, C., Feidt, M., Stanciu, D., Costea, M. (2015). Optimization of a vapor compression refrigeration system in transient regime. In *Advanced Thermodynamics of Irreversible Processes with Finite Speed and Finite Dimensions*, Petrescu, S., Costea, M., Feidt, M., Ganea, I., Boriaru, N. (eds). AGIR Publishing House, Bucharest.

Ust, Y. and Sahin, B. (2017). Performance optimization of irreversible refrigerator based on a new thermo-ecological criterion. *Appl. Therm. Eng.*, 30, 557–534.

Wijeysundera, M.E. (1999). Thermodynamic modelling of vapor absorption cooling systems. In *Recent Advances in Finite Time Thermodynamics*, Wu, C., Chen, L., Chen, J. (eds). Nova Science Publishers, New York.

4

Scientific and Technological Challenges of Thermal Compression Refrigerating Systems

Florine GIRAUD[1], **Romuald RULLIÈRE**[2] **and Jocelyn BONJOUR**[2]

[1] *Lafset, CNAM Paris, France*

[2] *CETHIL, INSA Lyon, Villeurbanne, France*

4.1. Introduction

All reverse cycle engines, based on cyclic transformation of a fluid, require the latter to undergo compression in part of the cycle. Thermal compression is the compression of vapors between two pressure stages using the physical phenomenon of sorption. Sorption is the binding of a gas, the "sorbate", corresponding to the working fluid in the context of energy processes, with a liquid or a solid, called the "sorbent". The sorbate and the sorbent form the sorption pair. A distinction is made between aBsorption, when the working fluid is in contact with a sorbent in liquid state, and aDsorption, when the working fluid is in contact with a sorbent in solid state. This physico-chemical reaction is exothermic.

The sorption reaction is reversible: a heat input allows the endothermic desorption of the working fluid contained in the sorbent.

Refrigerators, Heat Pumps and Reverse Cycle Engines,
coordinated by Jocelyn BONJOUR. © ISTE Ltd 2023.

If a comparison is made with mechanical compression, the sorption reaction corresponds to the suction phase of the working fluid in the compressor, while the reverse reaction, desorption, corresponds to its discharge.

NOTE.– In the case of sorption, the equilibrium being divariant, the equilibrium equations are of the form $p = f(T, x)$, linking pressure, temperature and mass fraction.

In order to work, these reverse cycles require contact with at least three heat sources: they are called "three-heat-source systems". Depending on the source temperatures used and the intended application, one sorption pair will be preferred to another, resulting in different scientific and technological issues.

The objective of this chapter is to give the reader the essential elements in order to understand the behavior of simple sorption systems, and to be aware of the scientific and technological issues currently raised by these technologies. Here, the most critical points when choosing and designing sorption systems are discussed, and the solutions usually implemented in currently marketed systems are described.

4.2. Kinetics and dynamics – heat and mass transfers in thermal compression engines

In addition to the intrinsic scientific challenges involved in understanding and describing the phenomena (sorption, heat transfer, advection-diffusion, thermodynamic non-equilibrium states, etc.), the coupling of the components in which the sorption and desorption reactions take place are numerous. The modeling of absorbers and adsorbers of reverse cycle engines is therefore always a delicate operation.

In this part, the reader will find the essential elements to consider the modeling of absorbers, then simple classically designed adsorbers, leaving it to them to ultimately tackle the modeling of more complex systems.

The heat and mass transfer phenomena occurring within the absorber or adsorber will be described and will make it possible to obtain a dynamic macroscopic representation of the behavior of these reactors and determine their thermodynamic performance.

4.2.1. *Absorption theory and design elements of absorbers*

4.2.1.1. *Absorption phenomenon*

Let us consider in this section the case of the absorption of superheated water vapor in contact with an absorbent solution composed of water and salt (generally LiBr; $CaCl_2$; $MgCl_2$; NaOH). The saline solution is subcooled, its water content is lower than that corresponding to the equilibrium state defined by the pressure of the system and the temperature of the solution. To return to equilibrium, the saline solution absorbs water vapor, resulting in water vapor condensation at the interface. The superheated water vapor condenses on a relatively hot interface. The driving force of absorption therefore corresponds to the distance from equilibrium, which can be expressed as the difference between the concentration at saturation (defined at the temperature of the solution and at the pressure of the system) and the actual concentration of water in the solution.

A number of phenomena appear during non-equilibrium reactions (Bird et al. 2007) including the Soret effect (Platten 2006), commonly called the thermodiffusion effect, and the Dufour effect (Ingle and Horne 1973), the opposite of the Soret effect. These two effects are, however, of less importance in the phenomenon of absorption. However, the Marangoni effect, due to the surface tension, and the inter-diffusion effect, which is linked to the binary nature of the absorbing solution, can play a significant role.

Surface tension gradients within the solution (resulting from temperature and concentration gradients) imply that liquid present in regions of low surface tension is pulled towards regions of higher surface tension (Carey 1989). This convective phenomenon, called the "Marangoni effect", induces instabilities and, if it exists in the saline solution, generates significant mixing near the interface, thus greatly improving heat and mass transfers.

In the case of absorption of water vapor by a lithium bromide solution, at the liquid/vapor interface, the LiBr concentration is lower and the temperature higher than in the core of the solution. However if this solution does not contain additives, the variations in surface tension as a function of temperature and concentration are very small (Kim et al. 1996), and tend to stabilize the liquid phase and cancel out the Marangoni effect (Nakoryakov et al. 2004).

However, the creation of a surface tension gradient can be induced by the addition of an additive or surfactant (generally alcohol hydroxides). Experiments were performed (Kashiwagi et al. 1985; Hozawa et al. 1991; Hihara and Saito 1992;

Ziegler and Grossman 1996; Nakoryakov et al. 2004, 2008) to study the effect of the active agent in the phenomenon of static absorption. 2-ethylhexanol (Yao et al. 1991; Kim et al. 1996; Kim and Infante Ferreira 2008) and n-octanol (Yao et al. 1991; Wu et al. 1998) are the most common in the literature.

Used at low concentrations (Kim et al. 1996; Nakoryakov et al. 2008), the additives make it possible to reverse changes in surface tension with temperature and lithium bromide concentration. Thus with the addition of additives, the surface tension increases with temperature and decreases with concentration, creating a gradient favoring the motion of the solution from the hottest zones (interface) towards the colder ones (center of liquid), and from the most concentrated (center of liquid) to the less concentrated (interface). The Marangoni effect thus ensures regeneration of the saline solution at the interface and promotes vapor absorption.

The inter-diffusion effect is linked to the binary nature of the absorbent solution (lithium bromide and water, for example). This phenomenon intervenes in the energy balance by considering that the interaction (diffusion of salinity) between a component, of enthalpy h_1 and the other component of enthalpy h_2, produces a significant additional amount of heat. This inter-diffusion effect is less than that of the Marangoni effect, but it is sometimes taken into account during numerical simulations in the equation where energy is in the form of a heat source.

During absorption, the thermophysical properties of the absorbent solution, which depend in particular on the concentration, vary. Taking this variation into account implies a depreciation of 5%–7% of the mass and thermal fluxes compared to the calculations performed with constant properties (Kawae et al. 1989; Shoushi et al. 2010).

4.2.1.2. *Falling film absorption model*

Many theoretical models of vapor absorption in a static absorbent solution have been developed (Grossman 1987; Bird et al. 2007), which describe the conservation equations in their most general form. They take into account the combined effects of mass and heat transfer. They are unsteady 2D models of convection/diffusion, based on Fick's law for mass transfer and Fourier's law for heat transfer.

The absorption of water vapor by a falling film constitutes a configuration that is encountered today in many absorbers (increasingly widespread in industrial applications). Unlike spray absorption, where water vapor and liquid interact on the surface of droplets, in falling film absorption, mass and heat transfer take place on the surface of a falling film, permeable to refrigerant. Absorption is governed by several parameters, including the flow regime and the physical properties of the fluid, in particular the mass and thermal diffusivities. In addition, the thermodynamic properties of the two fluids (temperature/pressure/composition) and the relationships between them considerably influence the transfers.

NOTE.– Below, the example of the absorption of water by a falling film of a solution of lithium bromide is considered. The proposed approach nevertheless remains valid for different sorption couples, as long as the chosen configuration concerns the absorption of a sorbate by a laminar film of sorbent with a high recirculation rate.

The absorption of water vapor by a solution of lithium bromide has been studied experimentally and numerically by the authors Nakoryakov and Grigor'eva (1977), Grossman (1983) and Kim and Ferreira (2008). The difficulty overcome during their studies lay both in the measurement and in the consideration of the thermophysical properties at the interface, where different phenomena occur: the precise estimation of the vapor mass flux, which diffuses in the film, is a real scientific challenge.

Figure 4.1 represents the hydrodynamic film (Zinet et al. 2012), the site of mass and heat transfer during absorption. On liquid inflow, the pressure and the concentration of the solution are lower than that of the vapor, thus creating a potential (gradient of pressure and concentration between the solution and the vapor). This potential promotes absorption, which results in a high solubility of the water vapor in the lithium bromide solution at the inlet of the absorber. Along the direction of flow, the solubility of water vapor in lithium bromide decreases due to the heat of absorption, which increases the temperature of the saline solution, reducing its absorption capacity. Also, heat transfers are faster than mass transfers.

The hydrodynamic film is governed by four main forces (Grossman 1986):

– the forces of gravity and inertia, due to the circulation of the liquid;

– the forces of viscosity or friction, which oppose the movement of the liquid at the wall;

– shear forces, due to the movement of water vapor at the interface;

– surface tension forces for thin falling films where capillary forces are dominant over gravitational forces. The surface tension forces generate instabilities at the interface, which changes from a smooth shape to a wave shape (Grossman 1986; Patnaik and Perez-Blanco 1996).

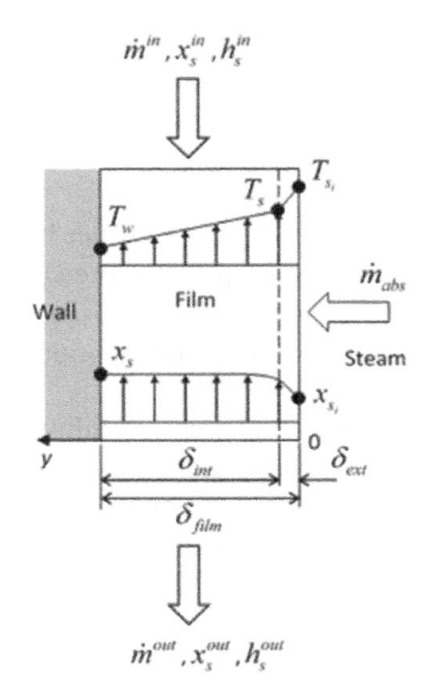

Figure 4.1. *Diagram of the modeled falling film specifying the temperature and LiBr concentration profiles proposed by Zinet et al. (2012)*

In the model developed by Zinet et al. (2012), the fluid is assumed to be perfectly mixed (homogeneous properties), the flow is assumed to be laminar, and the film is assumed to be rectangular and one-dimensional. These assumptions imply that the formation of preferential flow paths across the width of the absorber plates is not taken into account.

With these assumptions, the Nusselt stagnant film theory can be used to obtain the falling film thickness and velocity. The Reynolds number of the film is expressed as follows:

$$Re_{film} = \frac{4\dot{G}}{\eta_s} \quad [4.1]$$

where $\dot{G}$ is the mass flow rate per unit width of the film, and η_s is the dynamic viscosity of the solution. Several computational correlations of film thickness as a function of the Reynolds number have been proposed for laminar, transition and turbulent flow regimes (Grossman 1986). In many of them, a dimensionless "film thickness", called the Nusselt film thickness (Nu_T), is introduced. It is defined as follows:

$$Nu_T = \delta_{film} \left(g \frac{\rho_s^2}{\eta_s^2} \right)^{1/3} \quad [4.2]$$

with δ_{film} as the film thickness, g as the acceleration of gravity, and ρ_s as the density of the solution. According to the Nusselt theory, for the laminar regime, the thickness of the Nusselt film is related to the Reynolds number by the Brötz equation (Grossman 1986):

$$Nu_T = 0.0682\, Re^{2/3} \quad [4.3]$$

This correlation was defined to be in excellent agreement with experimental results over a wide range of Reynolds numbers, including transitional and turbulent regimes. Indeed, in falling films, even at high Reynolds numbers, a "laminar sub-layer" (relatively non-turbulent) occupies a significant part of the thickness of the film, whereas the turbulent effects occur mainly in the surface layer of the film. Thus, the laminar–turbulent transition is not well-defined, and the use of a correlation valid over a wide range of flow regimes, such as the Brötz equation, is more realistic.

Nusselt's analysis expresses the surface velocity of the film as follows:

$$v_{film} = \frac{3}{2} \frac{\dot{G}}{\rho_s\, \delta_{film}} \quad [4.4]$$

Considering that vapor absorption occurs only at the surface of the film, an exposure time based on surface velocity can be defined:

$$t_{expo} = \frac{L}{v_{film}} \quad [4.5]$$

where L is the film length. In the evapo-absorber, the length of the film is equal to the height of the plate. In the desorber, the length of the film is given by the outer half-circumference of the coil multiplied by the number of turns exposed to the vapor.

4.2.1.2.1. Mass transfer in the film

As shown by Auracher et al. (2008), the exposure time is much smaller than the time required for the concentration boundary layer to reach the wall. Thus, the mass transfer can be considered as the unsteady diffusion of water in a semi-infinite body and the concentration profile through the film thickness is given by:

$$\frac{c_{H2O}(y,t) - c_{H2O}(y,t=0)}{c_{H2O}(y=0,t) - c_{H2O}(y,t=0)} = erfc\left(\frac{y}{2\sqrt{D\,t}} \right) \quad [4.6]$$

where D is the diffusion coefficient of water in lithium bromide. Location $y = 0$ corresponds to the interface (Figure 4.1). The molar water concentration of the absorbent solution, c_{H2O}, is related to the mass fraction of lithium bromide x_s by the following relation:

$$c_{H2O}(y,t) = \frac{(1 - x_S(y,t))\,\rho_S}{\widetilde{M_{H2O}}} \quad [4.7]$$

with $\widetilde{M_{H2O}}$ as the molar mass of water.

The local mass flow rate of vapor absorbed into the film at a given location on the vapor/film interface ($y = 0$) can be expressed by Fick's law:

$$\dot{m}_{abs,local} = -S_{abs}\,\widetilde{M_{H2O}}\,D\left(\frac{\partial c_{H2O}}{\partial y}\right)_{y=0} \quad [4.8]$$

where S_{abs} is the absorption area. The total mass of vapor absorbed by the film between the inlet and the outlet of the absorption surface is obtained by integrating the local mass flow over the exposure time:

$$m_{abs} = \int_0^{t_{expo}} \dot{m}_{abs,local}\,dt \quad [4.9]$$

The mass of vapor absorbed can then be written as follows:

$$m_{abs} = S_{abs}(x_s - x_{si})\rho_S \frac{\sqrt{D\,t_{expo}}}{\sqrt{\pi}} \quad [4.10]$$

where x_{si} denotes the mass fraction of LiBr at the film interface, and x_s is the mass fraction of LiBr in most of the film. The assumption of ideal mixing at the film inlet implies a uniform inlet concentration across the thickness of the film:

$$x(y, t = 0) = x^{in} \quad [4.11]$$

Due to the very small amount of vapor absorbed during the exposure time, the concentration x_S remains equal to the concentration at the film inlet. However, the density of the solution in most of the film and at the interface can be considered equal to an average density ρ_S, as long as the change in concentration remains small.

The mean absorbed mass flow for the entire film is then written as:

$$\dot{m}_{abs} = \frac{m_{abs}}{t_{expo}} = \beta\,S_{abs}\,(x_s - x_{si}) \quad [4.12]$$

where β is the mass transfer coefficient:

$$\beta = \frac{2\,\rho_S}{\sqrt{\pi}} \sqrt{\frac{D}{t_{expo}}} \qquad [4.13]$$

4.2.1.2.2. Heat transfer in the film

Absorption is an exothermic process. As the vapor is absorbed at the film surface, the heat of absorption is transferred through the thickness of the film to the wall and the secondary heat transfer fluid, if present.

To address the heat transfer problem in components where cooling takes place, Flessner et al. (2009) extended the theoretical work of Auracher et al. (2008). In their work, the model becomes applicable for subcooled and superheated conditions in most of the film. To do this, the film is divided into two layers:

– the outer layer, in contact with the vapor phase, where the heat transfer between the surface of the film and the core of the film is coupled with the mass transfer of the absorption process;

– the inner layer, where the heat transfer between the core of the film and the wall is assumed to be independent of the mass transfer occurring at the surface.

In both layers, the temperature profile is assumed to be linear. However, in the outer layer, the temperature gradient is steeper than in the inner layer due to the local release of latent heat at the surface of the film. Using Nusselt's model, the heat transfer is calculated separately in each layer. The coupling between the two layers is obtained thanks to the interface temperature T_s. First, Flessner et al. (2009) took this temperature as equal to the temperature of the mixture at the film outlet, obtained from the energy balance over the entire film. They also assume the thickness of the outer layer is as follows:

$$\delta_{ext} = a\,\delta_{film} \qquad [4.14]$$

with the value of a fixed at 0.1. This constant should be determined from experimental data, but the value used by Flessner et al. (2009) at first approximation is consistent with other work in the literature (Jeong and Garimella 2002).

So, in the outer layer:

$$\dot{m}_{abs}\,\Delta h_{abs} = \alpha_{ext}\,S_{abs}\,(T_{si} - T_s) \qquad [4.15]$$

with Δh_{abs} as the latent heat of absorption, T_{si} as the surface temperature of the film, and α_{ext} as the heat transfer coefficient on the outer layer of the falling film. It should be noted that the coupling between heat and mass transfer is ensured in the model by equation [4.15].

In the inner layer, only a sensible heat transfer occurs. The heat flux (W) exchanged with the wall is written as:

$$\dot{Q} = \alpha_{int} \, S_{abs} \left(T_s - T_p\right) \qquad [4.16]$$

with T_p as the wall temperature. The heat transfer coefficient in each layer (external and internal) is:

$$\alpha_{layer} = \frac{k_S}{\delta_{layer}} \qquad [4.17]$$

where k_s is the thermal conductivity of the solution. Finally, the thermodynamic equilibrium between the vapor and liquid phases is assumed to exist only at the surface of the film:

$$T_{si} = T(x_{si}, p) \qquad [4.18]$$

where p is the pressure at the surface of the film.

4.2.2. *Adsorption theory and dimensioning elements of adsorbers and reverse cycle adsorption engines*

4.2.2.1. *Adsorption phenomenon: equilibrium and kinetics*

Adsorption can be physical in nature ("physisorption"), if the forces binding the adsorbate to the adsorbent are intermolecular (van Der Waals) or electrostatic. Adsorption is called chemical ("chemisorption") if a chemical reaction takes place between the adsorbate and the adsorbent (covalent force).

In the case of physisorption, by the action of intermolecular forces, the adsorption of adsorbate molecules on the adsorbent takes place gradually in the pores of the adsorbent until it covers the external surface of the pores. Since the bonding forces are strong between the adsorbate and the adsorbent, the adsorbate is in the solid state. Once the monolayer of adsorbent has been formed, the adsorbent attaches to this monolayer and then to the following layers until it forms a liquid layer in the pores of the adsorbent material. The phenomenon of fixation of the adsorbate on the adsorbent is exothermic and is characterized by a heat of adsorption, the value of which is of the same order of magnitude as the latent heat of vaporization of the chemical species considered. Conversely, the opposite phenomenon (desorption) is endothermic. It is by taking advantage of the absorption or release of heat that it is possible to use the adsorption phenomenon to achieve thermal compression, which will be the motive force of a reverse cycle engine.

The ability of an adsorbent to adsorb the adsorbate is the most determining factor for the performance of reverse cycle engines. This capacity is generally described by means of adsorption isotherms, curves parametrized by temperature and linking the quantity of gas adsorbed per unit mass of adsorbent (q^* for which the specific unit is used kg/kg_{ads}) to the pressure (p), often normalized by saturation pressure (p/p_{sat}), which makes it possible to describe the variation in pressure between 0 and 1. In general, increasing the pressure of the gas makes it possible to increase the amount of gas adsorbed (adsorbed phase concentration), while an increase in temperature reduces it (Figure 4.2). Several forms of mathematical relations have been proposed to describe isotherms (equations of Dubinin, Langmuir, O'Brien-Myers, etc.).

The mathematical form of the isotherms must be chosen with care for each adsorbate/adsorbent pair and for the range of operating conditions expected, because the accuracy of the values predicted by the model greatly affects the accuracy of the overall adsorber model.

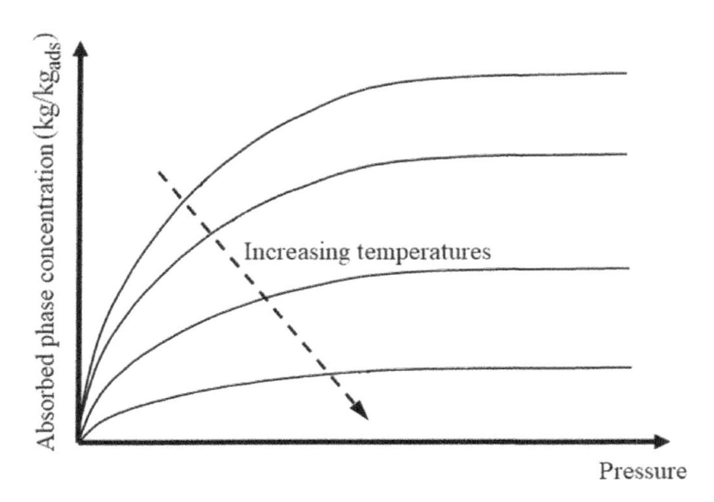

Figure 4.2. *Schematic representation of a set of adsorption isotherms*

Besides the description of the solid–gas thermodynamic equilibrium, the adsorption kinetics must also be taken into account. However, this kinetics can be limited by the following three phenomena:

– external mass transfer, which limits the movement of molecules through the boundary layers around the adsorbent material to get to its surface;

– internal mass transfer from the surface of the adsorbent material to its core. Transport through macropores and micropores is often distinguished, where the diffusion mechanisms differ (fluid phase diffusion, Knudsen diffusion, molecular diffusion, etc.);

– the intrinsic phenomenon of adsorption, which takes place at a finite rate. This last phenomenon is generally considered as non-limiting.

Each of these phenomena has been the subject of numerous studies and various models, and their competition governs the global kinetics which is, finally, often modeled by means of an adsorption coefficient (k_c) using the so-called Linear Driving Force (LDF) model, which is expressed mathematically in the form of a differential equation representing a mass balance:

$$\frac{\partial q}{\partial t} = k_c(q - q^*) \qquad [4.19]$$

with q as the adsorbed phase concentration and q^* as the concentration at equilibrium, as described by the adsorption isotherms. This adsorption coefficient is therefore specific to each gas–solid pair, and it can also depend on the geometric arrangements of the adsorber if the limit constituted by the external mass transfer is not negligible compared to the internal mass transfers. Various mathematical expressions (resulting from the combination of diffusion models) and coefficients adapted to different gas–solid couples (Table 4.1) can be found in the literature:

$$k_c = \frac{15D_{s0}}{R_p} exp\left(-\frac{E_a}{\bar{R}T_{ads}}\right) = k_1 exp\left(-\frac{k_2}{\bar{R}T_{ads}}\right) \qquad [4.20]$$

with $\bar{R}$ as the ideal gas constant, R_p as the pore radius of the adsorbent material, D_{s0} as a so-called asymptotic diffusion coefficient (obtained in the Henry domain), E_a as an activation energy, and T_{ads} as the temperature of the adsorbent.

	D_{s0} (m^2/s)	R_p (µm)	E_a (kJ/mol)	k_1 (1/s)	k_2 (K)
Zeolite 4A-water	1.31×10^{-9}	4,0	5.66	1228.12	680.7
Active carbon–methanol	N/A	N/A	8.13	7.35×10^{-3}	978
Silica gel type RD-water	2.54×10^{-4}	170	42.00	1.318×10^{-5}	5051.7

Table 4.1. *Some values of the parameters of the calculation of the global kinetic coefficient (Sharafian and Bahrami 2015)*

4.2.2.2. *Modeling of the adsorber of an adsorption-based refrigerator*

4.2.2.2.1. System description

In general, an adsorber can be considered as a reservoir containing the adsorbent material, exchanging heat with a heat transfer fluid that extracts heat during the desorption phase and rejects heat towards the heat transfer medium during the adsorption phase. Moreover, during the adsorption phase, the refrigerant in the form of vapor coming from the evaporator enters the adsorber where it is adsorbed. The cooling effect occurs in the evaporator. During the desorption phase, the refrigerant leaves the adsorber and enters the condenser where it condenses, which is accompanied by a release of heat, which creates the heating effect if the system is used as a heat pump. To increase the driving pressure differences, it is necessary to provide a pre-heating phase (before desorption, i.e. isosteric heating) and pre-cooling phase (before adsorption, i.e. isosteric cooling), during which the circulation of refrigerant is interrupted. Given this sequence of operations, an adsorption-based reverse cycle engine comprising only one adsorber can only produce cold (or supply heat) intermittently.

4.2.2.2.2. Material and energy balances

Here, a dynamic 0D model (at the scale of the components, i.e. each component is characterized by only one temperature) is presented to evaluate the time evolution of the temperatures of the various components and the mass of adsorbent present in the adsorber. Such a model makes it possible to roughly but quickly represent the behavior of the system and deduce its thermodynamic performances, in particular the thermal powers. This type of model can, of course, be improved by modeling the geometry of the components with more finesse, by including the calculation of pressure drops and their influence on the mass flow rates, and finally by taking into account thermal gradients (and possibly concentration in the adsorbed phase) in the different components.

Here, to maintain the generality of the subject, the heat exchanges with the heat transfer fluids are expressed directly in the form of thermal power deemed to be known and imposed or deduced as output variables of the model. These powers could be linked to the temperatures of the heat transfer fluids by using classic theories of heat exchangers or more advanced heat transfer models, depending on the very nature of the heat transfer system.

The energy balance of the adsorber, characterized by its temperature T_{ads} and the concentration of the adsorbed phase it contains q_{ads}, is written as:

$$\left(M_{ads}c_{p,ads} + M_{ads}q_{ads}c_{p,ff} + M_{met}c_{p,met}\right)\frac{dT_{ads}}{dt} = \dot{Q} + M_{ads}\left(\alpha c_{p,v}(T_e - T_{ads}) + \Delta h_{ads}\right)\frac{dq_{ads}}{dt} \quad [4.21]$$

with:

– M_{ads} as the mass of adsorbent in the adsorber and $c_{p,ads}$ as the specific heat capacity of the adsorbent;

– M_{met} as the mass of the metal parts making up the adsorber (generally essentially the heat exchanger it contains) and $c_{p,met}$ as the specific heat capacity of the metal used;

– Δh_{ads} as the specific heat of adsorption;

– $c_{p,ff}$ as the specific heat capacity of the liquid refrigerant and $c_{p,v}$ as that of the vapor refrigerant;

– the thermal power exchanged with the hot (desorption; $\dot{Q} > 0$) or cold source (adsorption; $\dot{Q} < 0$).

In the isosteric heating or cooling phase, the adsorbed phase accumulation term is eliminated (i.e. $\frac{dq_{ads}}{dt} = 0$) because the concentration in the adsorbed phase is practically constant: $q_{ads} = q_{min}$ (the lowest value encountered during the cycle, i.e. at the end of the previous desorption) during the cooling phase, and $q_{ads} = q_{max}$ (the highest value encountered during the cycle, i.e. at the end of the previous adsorption) during the heating phase. On the contrary, during the adsorption phase, $\frac{dq_{ads}}{dt} > 0$ and we also adopt $\alpha = 1$, while in the desorption phase, $\frac{dq_{ads}}{dt} < 0$ and $\alpha = 0$. Beyond isosteric phases, the term $\frac{dq_{ads}}{dt}$ is calculated using equations [4.19] and [4.20], retaining (necessary to calculate the adsorbed phase concentration at equilibrium q^*) the evaporation pressure for the adsorption phase, or the condensing pressure for the desorption phase.

The energy balance of the condenser, characterized by its mean temperature T_{cond}, is written as:

$$M_{cond}c_{p,cond}\frac{dT_{cond}}{dt} = \dot{Q}_{cond} - M_{ads}\left(c_{p,v}(T_{ads} - T_{cond}) + \Delta h_{lv(T_{cond})}\right)\frac{dq_{ads}}{dt} \quad [4.22]$$

with:

– $\Delta h_{lv(T_{cond})}$ as the latent heat of vaporization of the refrigerant at the temperature T_{cond};

– $\dot{Q}_{cond}$ as the thermal power released by the condenser ($\dot{Q}_{cond} < 0$);

– M_{cond} as the mass of the metal parts making up the condenser and $c_{p,cond}$ as the specific heat capacity of the metal used; the thermal inertia of the liquid contained in the condenser is disregarded with regards to the product $M_{cond}c_{p,cond}$.

This balance takes into account the thermal effects induced by the arrival of water vapor to the condenser from the adsorber, and then in the desorption phase characterized by the term of accumulation of phase adsorbed in the adsorber $\frac{dq_{ads}}{dt}$ (as pointed out previously, $\frac{dq_{ads}}{dt} < 0$ during this phase).

Finally, the energy balance of the evaporator, characterized by its average temperature T_{ev}, is written as:

$$M_{ev}c_{p,ev}\frac{dT_{ev}}{dt} = \dot{Q}_{ev} - M_{ads}\left(c_{p,ff}(T_{cond} - T_{ev}) + \Delta h_{lv(T_{ev})}\right)\frac{dq_{ads}}{dt} \qquad [4.23]$$

with:

– $\Delta h_{lv(T_{ev})}$ as the latent heat of vaporization of the refrigerant at the temperature T_{ev};

– $\dot{Q}_{ev}$ as the thermal power absorbed by the evaporator ($\dot{Q}_{ev} > 0$);

– M_{ev} as the mass of the metal parts making up the evaporator and $c_{p,ev}$ as the specific heat capacity of the metal used, with the thermal inertia of the liquid contained in the evaporator disregarded with regard to their product $M_{ev}c_{p,ev}$.

As for the condenser, this balance takes into account the thermal effects induced by the coupling between evaporator and adsorber: the adsorbed phase accumulation term in the adsorber $\frac{dq_{ads}}{dt}$ corresponds to the extraction of vapor from the evaporator to the adsorber, and then in the adsorption phase ($\frac{dq_{ads}}{dt} >$ during this phase).

4.2.3. *Issues associated with transfer kinetics and resistance*

4.2.3.1. *Contact time, residence time*

The affinity and reactivity of the sorbent with the working fluid constitute a chemical criterion on which the choice of the pair of fluids used in the sorption process must be based. It is this reactivity criterion that quantifies the sorption kinetics. However, from a macroscopic point of view at the scale of the absorber and the adsorber, this "kinetics", more often called "sorption rate", depends on the heat and mass transfers. In the quantification of the sorption rate, it is the transfer phenomena (in particular the mass transfer) which are generally limiting.

Most sorption models are based on the assumption that thermodynamic equilibrium prevails at the interface. The transfer of molecules across the interface occurs at a finite rate, and of all the molecules in contact with the surface, only those with enough energy to overcome the energy barrier can be ab- or adsorbed (Grossman 1986; Carey 1989). Therefore, only a fraction of the gas molecules can penetrate the liquid or be fixed in the pores in contact with the adsorbent. Upon penetrating the liquid, this fraction, which is expressed through the "accommodation coefficient", makes it possible to calculate the interfacial mass transfer coefficient. Thus, the overall resistance to mass transfer is the sum of the interfacial resistance and the resistance within the liquid. However, when the interfacial resistance is of the same order of magnitude as the resistance within the liquid, the assumption of a thermodynamic equilibrium at the interface is questionable.

The overall sorption rate can be quantified at the level of the sorption reactor from an equilibrium factor α defined by (Kim et al. 1995; Florides et al. 2003):

$$\alpha = \frac{x_o - x_i}{x_{o,eq} - x_i} \quad [4.24]$$

where $x_{o,eq}$ is the thermodynamic equilibrium mass fraction of the solution at its temperature at the outlet of the reactor and at the pressure in the reactor, and x_i and x_o are the actual mass fractions of the solution respectively at the inlet and at the outlet of the reactor.

Thus, mass fraction variations in the thickness of the solution film, and the impact of coupled heat and mass transfers in the film during its flow, can be taken into account. This parameter α makes it possible to represent the absorber in a simplified way within a global model of the absorption process.

4.2.3.2. *Adsorbers: cycle times and multi-bed strategies*

As explained previously (section 4.2.2.2.1), adsorption-based reverse cycle engines with only one adsorber operate discontinuously. The system is sized for a given cycle time (in particular, for solar systems where the cycle time is fixed by the day-night alternation), or it is the cooling load (or heating for a heat pump) that will allow the selection of the cycle time.

However, the adsorption and desorption phenomena within the adsorber are governed by different kinetics. In order to characterize a component or the adsorbents used, the adsorption and desorption dynamics of the component (respectively of the sorbent) are analyzed: the amount of working fluid adsorbed or desorbed over time is analyzed during adsorption/desorption cycles. These curves generally have exponential shapes like the curves obtained by Glaznev and Aristov

(2010) and presented in Figure 4.3 (curves obtained during the study of the isobaric adsorption and desorption dynamics of a single layer of silica gel on a metal plate).

The desorption dynamics presented in Figure 4.3 are 2.2–3.5 times faster than those obtained during the associated sorption phase (i.e. at the transition temperature of the cycle considered). This difference is due, beyond the specific characteristics of the sorbent/sorbate couple, to the difference in the conditions obtained during the adsorption and desorption phases: the desorption phase takes place at a higher temperature and vapor pressure than the adsorption phase. The dynamics are therefore more favorable during the desorption phases than during the adsorption phases. This difference should be taken into account when choosing the desorption and adsorption times of the installation.

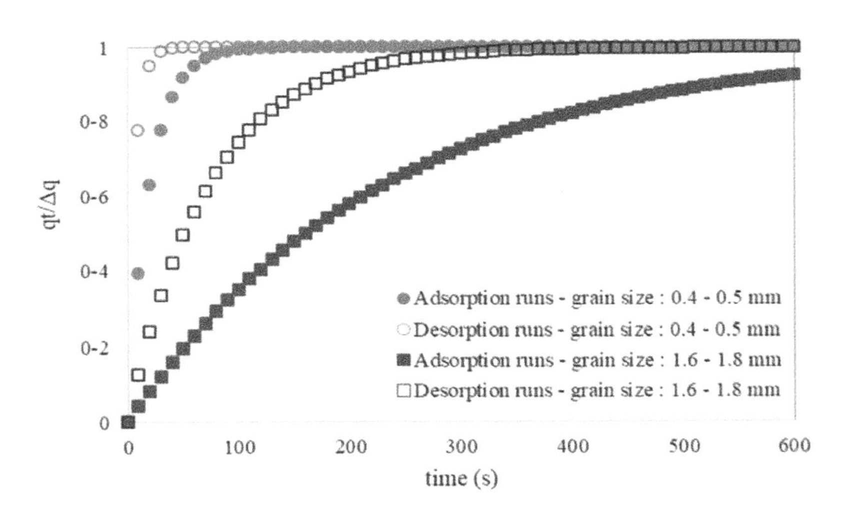

Figure 4.3. *Adsorption and desorption curves of water by silica gel for different conditions representative of the conditions obtained in adsorption engines (Glaznev and Aristov 2010). For a color version of this figure, see www.iste.co.uk/bonjour/refrigerators.zip*

Furthermore, for continuous cooling capacity, it is necessary to have at least two adsorbers, one operating in the adsorption phase while the other operates in the desorption phase, then reverse their roles – thus, the adsorption and desorption times are equal. But the interest of this configuration is then reduced by the fact that for about two-thirds of the cycle time, the adsorber in the desorption phase is in contact with the hot source without being able to be significantly regenerated. According to the heat rejected, the temperature and pressure levels, and the adsorption or desorption kinetics, we can then optimize the adsorption (by accepting that the adsorber is never used at its maximum theoretical capacity) and desorption time to meet a given constraint (maximization of the COP, maximization of the cooling capacity, etc.), as shown theoretically by Miyazaki et al. (2009). Finally, at the cost of greater complexity, it is possible to move towards using the adsorber at its maximum theoretical capacity, provided that three adsorbers are used, and by optimizing the distribution of adsorption and desorption times (Zajaczkowski 2016).

4.3. Technological challenges in component design

4.3.1. *Fluid pair*

Currently, the most commonly used refrigerant/sorbent pairs in absorption systems are ammonia/water and water/lithium bromide (LiBr). This observation was made, in particular, by Wang et al. (2015), who carried out a state of the art of absorption systems based on 109 experimental studies (see Figure 4.4).

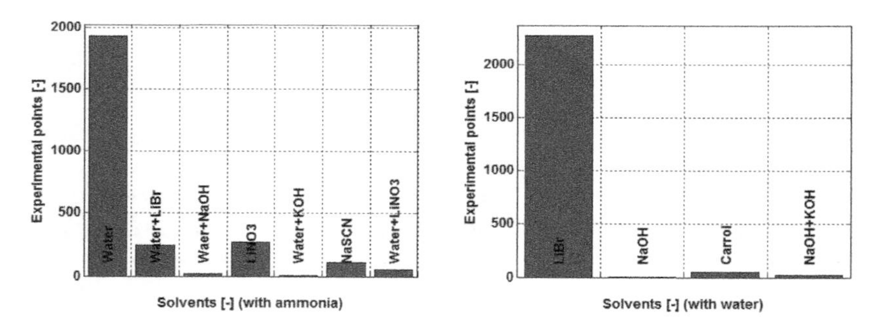

Figure 4.4. *Count of experimental points in the literature specifying the main sorbents used with ammonia (left) and with water (right)) (Wang et al. 2015). For a color version of this figure, see www.iste.co.uk/bonjour/refrigerators.zip*

In the ammonia/water couple, water plays the role of the absorbent solution and ammonia plays the role of the refrigerant. Ammonia has a high affinity with water and has a very low solidification temperature (-77°C). This pair therefore makes it possible to reach temperatures below 0°C and is therefore suitable for freezing applications. Ammonia is neutral with respect to the ozone layer and has no global warming potential (GWP). However, it is flammable and toxic. In addition, its vaporization temperature is close to that of water at high pressure, hence the possibility of finding small amounts of water in the ammonia after regeneration (desorption). A rectifier is therefore necessary to improve the separation of the refrigerant from the sorbent.

Finally, ammonia attacks copper and its alloys. Refrigeration installations must therefore be made of stainless steel or steel with the addition of corrosion inhibitors.

In the water/lithium bromide pair, water plays the role of refrigerant and lithium bromide plays the role of absorbent solution. This pair has a great affinity. The boiling point of lithium bromide is much higher than that of water, hence the absence of a refrigerant/solution mixture at the desorber outlet. The use of water has many advantages: very abundant fluid in its natural state and inexpensive, it is non-toxic and non-flammable. Water has zero global warming potential and is neutral with respect to the ozone layer. LiBr is not toxic and its GWP is equal to 0. However, the use of this couple, without adding components in the refrigerant (see section 4.4.2), does not allow working at lower temperatures at 0°C, due to the transformation of water into ice. The water/lithium bromide pair is therefore suitable for applications that do not require negative temperatures, for example, air conditioning. The crystallization of the saline solution, which appears for high concentrations, must be closely monitored during its use (see section 4.4.1). The pressure inside a system operating with the water/lithium bromide pair is approximately 50 times lower than atmospheric pressure and this generates technological constraints to ensure the leak-tightness of the components. Maintaining a low level of pressure is essential to the efficiency of the system because the presence of non-absorbable gases considerably reduces its performance.

Regarding adsorption chillers or heat pumps, the systems currently marketed mainly use water or ammonia as the refrigerant. Ammonia is adsorbed by activated carbon or salts, whereas water is generally used with zeolite or silica gel; zeolite generally requires a higher regeneration temperature than silica gel (above 70°C except for zeolites such as SAPO-34, which can be regenerated at temperatures of the same order as those used for silica gel). Problems similar to those mentioned for the couples used in absorption are then observed, in particular the need to comply with the regulations related to the use of ammonia and combat corrosion phenomena if ammonia is used and the problems related to maintaining a primary vacuum and freezing if water is used.

4.3.2. *Absorber*

The absorber is a heat and mass exchanger. As it is the site of the sorption phenomenon, it is generally a key component in the overall process, and its design, characterization and optimization are important topics.

There are multiple challenges in its design: in particular, it is a question of allowing the sorption phenomenon (which implies facilitating the access of the liquid-vapor interface to the gaseous working fluid) to efficiently transfer the absorbed heat, while limiting the weight and volume of the system, as well as its cost.

Overall, there are three main factors to improve absorption performance:

– the temperature (when the temperature of the absorbent solution decreases, the solubility of the sorbate in the sorbent generally increases);

– the contact surface between the vapor and liquid phases;

– the contact time between the phases.

The different types of absorbers frequently used in the industry are:

– packed tower;

– tray tower;

– bubble tower (with or without mechanical agitation);

– falling film tower (vertical or horizontal);

– sprinkler or spray towers;

– Venturi or ejector absorbers.

Depending on the technology of the reactor, the contact surface between the vapor and the absorbent solution is generated by different flow configurations: the liquid can flow on a support and form a thin film, the liquid can flow in the gas phase in the form of droplets or the gas can flow in the form of bubbles within the liquid itself.

The compactness of the systems has so far been little optimized (these systems are used in industrial processes, with, for example, exchangers of the shell and tubes type with the solution flowing outside the horizontal tubes (Killion and Garimella 2001)); however, this criterion is becoming more important today in the context of domestic use, for example.

In refrigeration applications, the choice of absorber is generally that of falling film columns (Keizer 1982). In these processes, absorption takes place on the surface of a film dripping from the solution either outside (Kim et al. 1995) or inside (Takamatsu et al. 2003) vertical smooth tubes, or on smooth horizontal tubes (Kyung et al. 2007; Kang et al. 2008) (Figure 4.5).

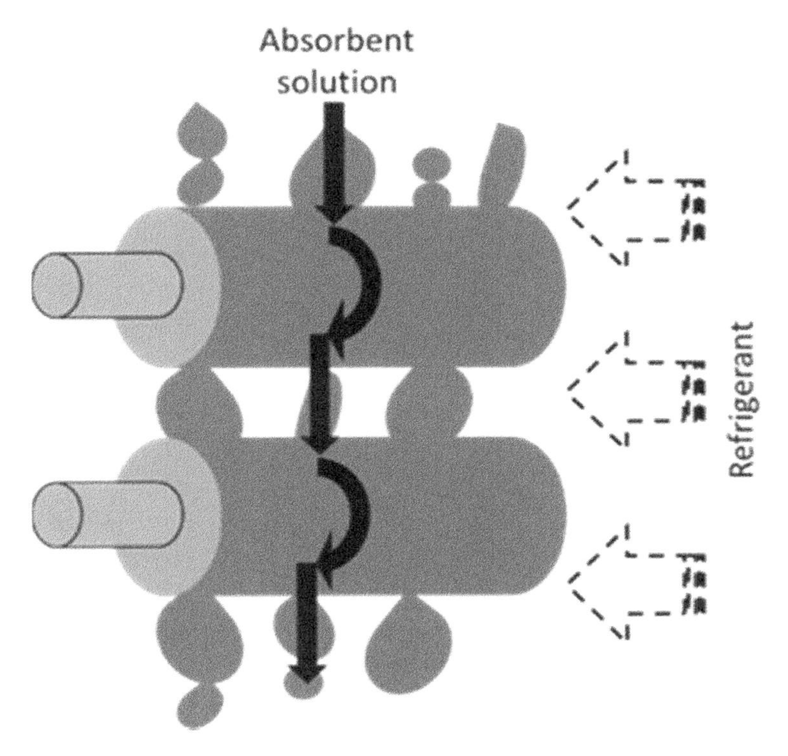

Figure 4.5. *Schematic representation of a horizontal tube absorber. For a color version of this figure, see www.iste.co.uk/bonjour/refrigerators.zip*

Of the falling film absorbers, the tubular horizontal absorber is the most common since it allows homogeneity of the distribution of the liquid and has a certain ease of installation (Hu and Jacobi 1996). With this configuration, a thin film of liquid solution emerges from a fluid distributor, in the upper part of the absorber, and flows downwards by gravity over a group of horizontal tubes. The tubes are surrounded by the refrigerant in vapor state coming from the evaporator. During absorption, heat is released at the liquid–vapor interface. If this heat is not removed or transferred, the temperature of the liquid increases and the absorption slows down. To transfer the released absorption heat, a heat transfer fluid circulates inside the tubes. The performance of an absorber is influenced by the concentration of the solution, the flow velocity, the absorption temperature and the nature of the tube surface.

Many studies have shown that the absorption phenomenon is controlled by the transfer resistance (in particular, mass transfer) in the liquid phase. Indeed, the working fluid which is absorbed at the liquid–vapor interface is in general not mixed efficiently in the liquid. It therefore remains close to the interface and tends to slow down the absorption of additional vapor. Many experiments have been carried out with the aim of increasing the heat and especially mass transfer coefficients (Kim et al. 1995; Deng and Ma 1999; Arzoz et al. 2005; Chengming et al. 2010). Most of the work reveals that additives increase the heat and mass transfer coefficients through a combined effect of surface tension reduction, as well as an increase in the Marangoni effect, which induces convection currents (turbulence at the interface), promoting absorption. Improved transfer comes from the shape of the interface generated by these surface tension gradients. Indeed, the interface is no longer a smooth film but a wavy film.

Two types of film have been distinguished by Patnaik and Perez-Blanco (1996) and adapted by Killion and Garimella (2001) depending on the flow regime (transition or turbulent regime):

– the case of a capillary wave interface which appears in the transition regime, i.e. for $20 < Re < 200$. In this case, the waves have a low amplitude but a higher frequency. This is the most common case in the literature;

– the case of a roll wave shaped interface. In this case, the amplitude of the waves is greater but with a lower frequency. The formation of these waves induces convection movements that intensify the transfers. This type of interface is formed in a turbulent regime, i.e. for $200 < Re < 4{,}000$.

For small solution flow rates, it can be observed that the wall is not completely wetted because the falling film tends to break and form rivulets that do not cover the entire surface. The minimum wetting rate, beyond which complete wetting of the surface is possible, can be determined theoretically from energy considerations (Bankoff 1971). Additives can be used to improve wetting by promoting the Marangoni effect, which tends to bring turbulence to the interface and improve mass transfers. Kim and Infante Ferreira (2008), in addition to studying the effect of additives, attempted to improve wetting by attaching a metal grid to the plate. This grid aims to spread the liquid over the entire plate by capillarity. The results of their tests show that, without additives, the grid improves mass transfer by promoting wetting of the plate. However, the effect generated by the addition of additives (strong Marangoni convection) is reduced by the use of a grid, which stabilizes the liquid by capillarity. The study conducted by Kim and Infante Ferreira (2008) also shows that, during absorption, the mass transfer is increased when the absorbent solution is cooled. This conclusion is in line in particular with the analytical calculations of Grossman (1983) and Brauner (1991). The heat released by absorption at the liquid–vapor interface (exothermic reaction) must be transferred through the liquid phase to the wall of the exchanger. Indeed, an increase in interface temperature leads to a modification of the operating conditions and therefore also tends to slow down the rate of sorption. In addition, heat transfers on the side of the external heat transfer fluid can also be limiting (Herold et al. 1996) and a good overall design of the reactor is therefore essential for the proper functioning of the system.

A point of vigilance in trickle-bed reactors concerns the quality of the distribution of the fluid flow on the walls of the tubes/plates. The efficiency of the absorber is greatly influenced by the thickness of the film. Improper distribution can cause some tubes/plates to become flooded and others to dry out. In addition, the formation of rivulets may appear, in particular when the reactor is operating at a low solution flow rate, also leading to the drying out of certain areas of the wall. An additional constraint imposed on these distributors is to present very low-pressure drops and to be able to operate over a wide range of operating conditions.

Finally, the presence of non-absorbable gases within the absorber (related to sealing problems or caused by corrosion reactions) is responsible for a significant reduction in the absorption rate. These gases are driven towards the liquid–vapor interface and tend to accumulate near it. Therefore, water vapor must diffuse through the non-absorbable layer to reach the interface, and this resistance to mass transfer greatly reduces the absorption rate. In the presence of 10% non-absorbable gas at the interface, Kim et al. (1995) show that the mass transfer rate is half that of the case where there are no non-absorbable gases.

4.3.3. *Adsorber*

The adsorber has long been considered the critical component, which through its optimization, improves the cost and compactness of adsorption systems. As explained in section 4.2.3.2, this component is alternately the site of an adsorption and then a desorption reaction. Although adsorption kinetics can be up to 3.5 times slower than desorption kinetics, it is for this phase that researchers generally try to improve the technical characteristics of the component. Several ways have been identified in order to design effective adsorbers. Beyond the characteristics specific to the adsorbent itself (good mechanical and thermal stability, low ageing, ability to not release inert gases during the various cycles, etc.) and to the adsorbent/working fluid pair (adsorption capacity, heat of adsorption, etc.), an effective adsorber must have the following characteristics, particularly:

– a high heat transfer coefficient between the exchanger (in particular, the secondary surfaces of the exchangers) and the adsorbent;

– a high mass transfer coefficient (inter- and intra-particle);

– a low "exchanger/adsorbent" mass ratio.

Indeed, the first two characteristics (high mass and heat transfer coefficients) make it possible to obtain a high specific power by quickly reaching the rate of 80–90% of the total adsorption capacity of the adsorbent (rate usually considered optimal for calculating cycle times). The third characteristic (exchanger/adsorbent mass ratio), despite having significant repercussions on the dynamics of adsorption, makes it possible to improve the coefficient of performance by reducing the fraction of material heated during sorption/desorption cycles.

Besides these considerations, there are also constraints in terms of the choice of materials making up the exchanger (corrosion, durability, low heat capacity, high thermal conductivity), as well as sealing constraints. The presence of inert gases greatly degrades the performance of the installation by creating resistance to additional transfers (of mass and heat) and by modifying the operating conditions (in particular, by adding an additional partial pressure). From these constraints, several adsorber technologies have been developed.

4.3.3.1. *Granular adsorbers*

The simplest adsorber consists of an exchanger, generally with fins, readily available on the market and developed for applications such as air conditioning or the cooling of heat-dissipating systems (engines, etc.). The most commonly used exchangers in sorption systems are finned tube and plate-fin exchangers. In general, an exchanger with an exchange surface of 500 m^2/m^3 can be favorably used as an exchanger in adsorbers (Schnabel et al. 2018). The space available between the fins is filled with adsorbent and generally surrounded by a metal net in order to maintain the grains of adsorbent in place. The exchanger filled with adsorbent is then placed in a sealed box. An example of an adsorber designed using this technique is shown in Figure 4.6.

This adsorber design offers the advantage of being inexpensive and easy to operate. Energy densities of 50–100 $W.dm^{-3}.K^{-1}$ can be obtained (Schnabel et al. 2018). However, this configuration has several disadvantages, in particular low heat and mass transfer and a high exchanger/adsorbent mass ratio (due, in particular, to the addition of an additional metal net to retain the adsorbents). To limit these drawbacks, studies have been carried out to study the impact of various parameters, including the geometric characteristics of the exchangers and the fins, as well as the size and arrangement of the grains.

Figure 4.6. *Example of an adsorber made from a finned tube heat exchanger in the process of being filled. For a color version of this figure, see www.iste.co.uk/bonjour/refrigerators.zip*

4.3.3.1.1. Geometric characteristics and density of fins

In order to guarantee good adsorption and desorption, the temperature of the adsorbent bed must be as homogeneous as possible. This condition makes it possible, among other things, to prevent vapor from being desorbed in one part of the adsorber to be re-adsorbed in another. However, too high a fin density leads to an increase in the exchanger/adsorbent mass ratio and consequently to a reduction in the overall coefficient of performance (COP, see Chapter 1) of the system. Furthermore, a too small inter-fin space can also lead to less efficient mass transfer by reducing the cross section of the vapor and by creating greater pressure drops. The specific power is therefore reduced while it tends to increase with the reduction of the fin pitch. In order to increase the space between the fins, it is then possible to play on the fin size and the geometry: a greater fin height makes it possible to increase the inter-fin space. To optimize the dimensions of the fins, the aim is to optimize the fin aspect ratio (height/pitch). The results of the studies carried out on the impact of these characteristics generally led to recommending an inter-fin spacing of the order of 6 mm (between 5 and 7 mm). This spacing is to be adjusted depending on the geometry of the exchanger and the size of the adsorbent particles. For a given vapor path length, the higher the efficiency of the exchanger, the more the inter-fin spacing can be increased. For example, in the study by Golparvar et al. (2018), the inter-fin spacings recommended by the authors for the intended application and for zeolite-13X (particle diameter of 0.4 mm) are 5.4 mm for a heat exchanger with longitudinal fins and 6.8 mm for an exchanger with annular fins, for a fin height of 10 mm (respectively 5.0 mm and 6.3 mm if the fin height is 20 mm). This difference is due to the thermal characteristics of the exchanger, which make it possible to obtain a uniform temperature field more quickly in the case of annular fins than in the case of longitudinal fins. Regarding the grain size, the smaller the size of the adsorbent particles, the narrower the space available for the particles must be and conversely, the larger the size of the adsorbent particles, the larger the available space must be, in order to optimize the specific power (Mitra et al. 2018).

4.3.3.1.2. Grain size

The grain size is a parameter that significantly influences the adsorption dynamics. Mass transfer in the adsorbent bed is controlled by intra-particle (circulation of vapor through the space left between the particles) and inter-particle (circulation in the vapor within the particle itself) diffusion. A small grain size will tend to increase the resistance to inter-particle transfer, while a large grain size will tend to increase the resistance to intra-particle transfer. The advantage of using small grains is that the transfer of mass and heat in the grains is all the more reduced the smaller the particles are. In addition, the exchange surface provided by the adsorbent for a given mass of adsorbent is increased. However, as seen previously, choosing small grains implies increasing the number of fins so as to not to increase the number of layers too much (which would increase resistance to mass heat transfer). This increase in the number of fins leads to an increase in the exchanger/adsorbent mass ratio, and therefore to a reduction in the COP. Furthermore, from a technical point of view, choosing very small grains involves using a net to hold the grains in place, made up of very fine meshes, which increases resistance to the transfer of vapor when it enters the adsorber. Aristov et al. (2012), however, showed that there is a particle size for which the adsorption dynamics are no longer dependent on the grain size, since below this size, the heat transfer drives the dynamics, since the resistances to mass transfer have become very weak with regard to resistances to heat transfer. In this study, the grain size for which this transition is observed is 0.8 mm (study carried out with silica gel). The same transition is observed by Santamaria et al. (2014) with zeolite (AQSOA Z02) for a grain size of 0.7 mm. However, the authors note that for a grain size below 0.3 mm, the resistance to intra-grain transfer becomes limiting again, significantly deteriorating the sorption dynamics (unless a flat plate is used). Thus, the grain size must be within a given range and does not necessarily have to be homogeneous.

As an indication, Table 4.2 lists a few adsorbent diameters studied in the scientific literature with, when a conclusion can be given, the diameter with the best performance in terms of adsorption dynamics.

Couple	Range tested (mm)	d opt (mm)	Reference
Water/Fuji silica gel	0.2–1.8	< 0.8	Aristov et al. (2012)
Water/silica gel	0.1–0.8	0. 2	Niazmand and Dabzadeh (2012)
Water/zeolite (AQSOA Z02)	0.15–1.18	0.3–0.7	Santamaria et al. (2014)
Water/zeolite (FAM-Z02)	0.7–2.6	0.7–1	Dawoud (2007)
Water/FAM-Z01 (zeolite)	0.01–0.3	0.05	Duong et al. (2020)
Ethanol/activated carbon (Maxsorb III)	0.03 and 0.07		Mitra et al. (2018)
Water/SWS-1L	0.1–0.8	0.2–0.3	Niazmand et al. (2013)

Table 4.2. *Example of the diameter range of adsorbents studied in the scientific literature and, if known, the diameter range recommended by the authors*

Beyond the grain size, the adsorption dynamics is also impacted by the arrangement of the adsorbent layers (porosity). If a large porosity is favorable to the mass transfer, it, on the contrary, is unfavorable to the heat transfer. However, a significant part of the heat transfer resistance is due to the contact resistance at the metal/adsorbent contact level. Achieving a quality metal/adsorbent contact so as to not deteriorate the heat transfer efficiency is a real challenge. These observations have gradually led to the development of adsorbers with coated surfaces: the adsorbent is no longer in the form of grains but in the form of powder, which is then deposited around the heat exchanger.

4.3.3.2. *Adsorbers with coated surfaces*

The development of heat exchangers with coated surfaces has made it possible to considerably reduce resistance to mass and heat transfer. The specific power of adsorbers per kilogram of absorbent could thus be increased by a factor of approximately 10 (Meunier 2016). In these exchangers, the adsorbent is no longer in the form of grains, but in the form of powder attached to the exchanger. There are mainly two techniques for coating the adsorbent powder on the exchanger: direct crystallization (the coating is created directly on the surface of the exchanger) or the use of binders (the powder is maintained on the exchanger and between it thanks to a binder). Examples of exchangers with coated surfaces are shown in Figure 4.7.

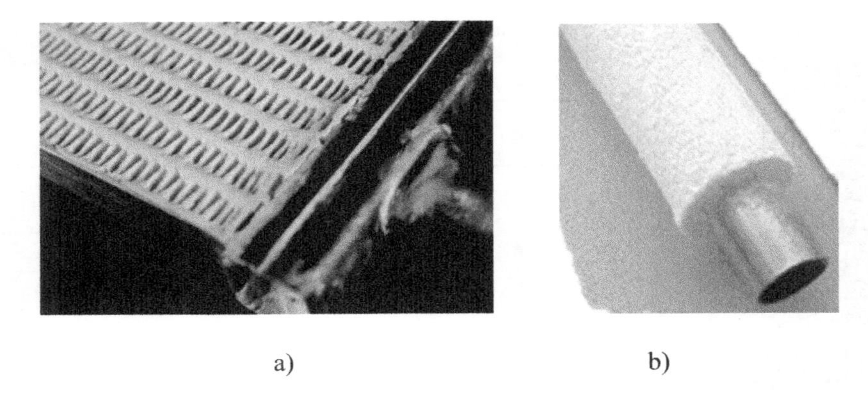

a) b)

Figure 4.7. *Example of coated surface heat exchanger presented in a) Bendix et al. (2017), b) Restuccia et al. (2004)*

Compared to granular adsorbers, coated adsorbers generally make it possible to accelerate mass transfer by limiting intra- and inter-grain transfer resistances, by reducing thermal resistance since the thermal conductivity of the layer is increased and by reducing the exchanger/adsorbent contact resistance. These improvements in mass and heat transfer allow a reduction in the cycle time and an increase in the power density.

4.3.3.2.1. Impact of the coating technique used

Several direct crystallization techniques are possible. The coating technique by direct crystallization, based on hygrothermal synthesis, roughly consists of immersing the surface to be coated in a reactive solution and then heating it at high temperature for several hours. The surface is then washed and dried. The reaction is mainly driven by pressure and temperature. The created layer is regular and homogeneous even for complex geometries. It is possible to adapt the morphology of the layer to the intended application.

Coating by crystallization has the advantage of combining, in one step, the manufacture of the layer, as well as its deposition on the exchanger. It also has the advantage of considerably reducing the contact resistance between the metal and the adsorbent layer, which thus becomes very low compared to the contact resistances obtained with granular beds or by using a binder. However, the process is quite complex to implement and requires high pressures and high temperatures. It is therefore quite expensive.

The coating technique, using an organic or inorganic binder, consists of creating a homogeneous aqueous solution of adsorbent and binder. Then, the solution precipitates by various processes, forming a gel, which is deposited in a layer on the metal of the exchanger. Heat treatment then stabilizes the layer.

This technique is easily used on a large scale and for exchangers composed of different adsorbents and metals. The major difficulty lies in the choice of the binder: it must have interesting thermal characteristics, but must not obstruct the pores of the adsorbent powder nor degrade over time (releases of incondensable gas have been observed following degradation of the binder). However, the coating layer created is more porous than that obtained by crystallization, which allows the working fluid to more easily access the adsorbent. The energy density obtained with this technique is therefore generally higher.

The two techniques presented above lead to similar thermal conductivities of the layer. For reasons related to mechanical resistance, they are also both limited in terms of achievable film thickness (typically the equivalent of 150–300 kg of adsorbent per cubic meter of exchanger – Calabrese et al. (2019)). As mentioned above, on the contrary, the contact resistance is lower if the crystallization technique is used, but the energy density obtained is generally lower.

Both techniques are used in commercial facilities (Sortech AG for crystallization, Mitsubishi Chemical Corporation, Nanoscape GmbH, or Klingenburg GmbH for beds with binders – Schnabel et al. 2018).

4.3.3.2.2. Layer thickness

In general, it is preferable to have the thinnest possible coating layer in order to limit resistance to heat and mass transfer. In addition, the greater the thickness, the closer the performance of the adsorber to the performance obtained with granular adsorbers (in terms of adsorption dynamics). As an example, Dawoud (2013) notes an improvement in adsorption dynamics of only 7% with a coating thickness of 500 µm compared to a granular exchanger (FAM-Z02 grain sizes between 0.7 and 1 mm), but doubles or even triples the adsorption rate with a layer thickness of 200 µm (zeolite FAM-Z02). However, the thickness of the adsorbent layer does not only have an impact on the specific power of the installation; it also has a significant impact on the price (a significant part of the final cost of the installation is due to the cost of adsorbents) and the coefficient of performance of the installation. If a thin layer makes it possible to reduce the cost of the installation and obtain rapid adsorption dynamics (and therefore a high instantaneous power), with the exchanger/adsorbent mass ratio being high, the coefficient of performance of the installation is greatly degraded (the coefficient of performance varies practically linearly with the thickness of the coating). Thus, as Dawoud summarized:

> [...] the design engineer has to choose between accepting the reduction in COP to obtain the required thermal power output, or compromising on the thermal power output to get the highest COP, despite the need for more zeolite [adsorbent] mass and the related higher cost (Dawoud 2013, p. 1650).

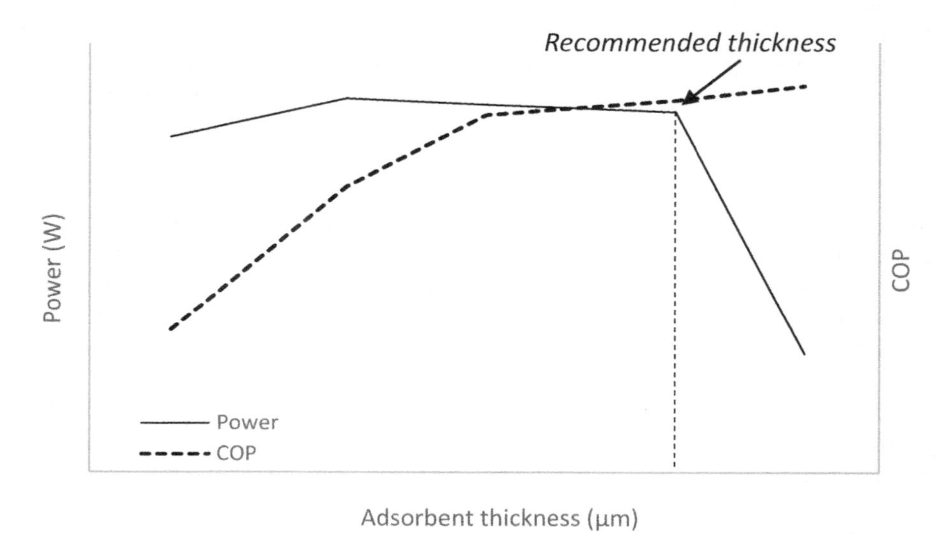

Figure 4.8. *Example of evolution of power and COP versus thickness of adsorbent coating observed in the scientific literature (Bendix et al. 2017; Duong et al. 2020)*

However, studies have shown that there is a thickness that allows significant thermal power to be obtained while maintaining a slightly degraded coefficient of performance. Bendix et al. (2017) showed that, by studying, for a given exchanger, the impact of the thickness of the adsorbent powder (zeolite – TiAPSO), it is possible to increase the coating thickness while maintaining a high thermal power (the cycle time is adapted to the amount of adsorbent coated on the exchanger in order to obtain, for each thickness, 90% of the adsorbed working fluid load). A plateau is then observed (Figure 4.8). The recommended adsorbent thickness would therefore be the thickness at the end of the plateau. It should, however, be noted that the ideal exchanger/adsorbent mass ratio considered by Bendix et al. (2017) (ratio of

1.04 or 313 kg/m^3 – adsorbent: TiAPSO zeolite) is much lower than that considered in most of the studies carried out on this topic (see Table 4.2).

Pair	Coating technique	kg_{HEX}/kg_{ad} ratio	Adsorbent density kg_{ad}/m_3	Optimum kg_{HEX}/kg_{ad} ratio	Reference
		Thickness µm		Optimal thickness µm	
Water/TiAPSO (zeolite)	Binder	0.85–7.42	382–44	1.04 (313 kg_{ad}/m_3)	Bendix et al. (2017)
		NA		NA	
Water/AQSOA-Z02 (zeolite)	Binder	2.59–4.37	168–97	2.59	Dawoud (2013)
		500–150		200	
Water/SWS-8L	Binder	3.8–2.2	226	NA	Freni et al. (2012)
		NA			
Water/FAM-Z01 (zeolite)	Binder (numerical study)	NA	NA	NA	Duong et al. (2020)
		500–50		150	

Table 4.3. *Example of coating thickness studied in the scientific literature and, if known, exchanger/adsorbent mass ratio or thickness recommended by the authors*

The thickness of the layers generally deposited in coated exchangers is 150 µm or even a little more (Meunier 2016). As an indication, Calabrese et al. (2019) estimate that the threshold value for zeolite (SAPO-34) to allow acceptable adsorption performance is 135 µm. The exchanger/adsorbent mass ratio is therefore higher for coated surfaces: the specific power is increased, but not the coefficient of performance. In order to reduce this ratio, adsorbers in which the fins are replaced by metal foams, and on which the direct crystallization of adsorbent is carried out, are under research and development.

4.3.4. *Evaporator*

Depending on the pairs chosen, the design of an efficient evaporator can prove to be a more or less easy task. For fluids whose operating pressure and temperature ranges in sorption systems are similar to those used in other applications, the design of the evaporator itself does not pose any real difficulty. This is the case, for example, for systems using ammonia as the refrigerant for which the evaporators marketed, for example, for refrigeration (shell and tube, plates), may be suitable for sorption systems. However, if water is used as the refrigerant, which is the case in many commercial installations, in particular for single-effect absorption water chillers (Altamirano et al. 2019), the operating conditions pose real technical difficulties. Indeed, depending on the intended application conditions and the temperature, the pressure within the evaporator can be up to 100 times lower than atmospheric pressure.

NOTE.– The water saturation pressure at 10°C – which typically corresponds to operating conditions in a heat pump or for air cooling – is 1.2 kPa (12 mbar). At such pressures, the weight of the water column results in a static pressure of the same order of magnitude as the saturation pressure, which induces remarkable thermodynamic effects, in particular local subcooling of the water.

Thus, if the exchanger technology induces the presence of a water column (case of flooded evaporators, for example), the pressure variation due to the hydrostatic pressure within the column must be taken into account. As the saturation temperature varies with pressure, the greater the hydrostatic pressure, the higher the fluid saturation temperature (Figure 4.9). It is therefore necessary to reach a higher temperature within the fluid at the bottom of the water column to initiate the phase change than at the free interface. Consequently, if the secondary fluid enters the evaporator at a temperature of 20°C, considering the example presented in Figure 4.9 and assuming a pinch of 5°C, it is no longer possible to initiate the phase change from a water column depth of about 5 cm. The energy density of the exchanger is therefore very greatly degraded, especially since this phenomenon adds to the difficulty of triggering boiling due to the increase in the theoretical superheating necessary and a decrease the number of nucleation sites potentially activatable with the decrease in pressure.

Indeed, at these pressures, the density of the vapor is very low. As an indication, it is 0.009 kg/m^3 at a saturation temperature of 10°C, i.e. 64 times lower than that obtained at atmospheric pressure. However, the theoretical superheat (ΔT_{sat}) required for nucleation for a given nucleation site size (R_{cav}) (formula derived from the Clapeyron formula – equation [4.25] – with T_w being the boundary temperature, σ being the surface tension of water, ρ_v being the density of the vapor, and Δh_{lv} being the latent heat of vaporization), and the size of the sites that can be activated, are dependent on the thermophysical properties of the fluid, in particular the density. The size of potentially activatable sites also depends on the saturation temperature (Hsu 1962). Thus, the lower the density and the saturation temperature, the greater the theoretical superheat to trigger boiling (heterogeneous nucleation) and the lower the number of sites that can potentially be activated.

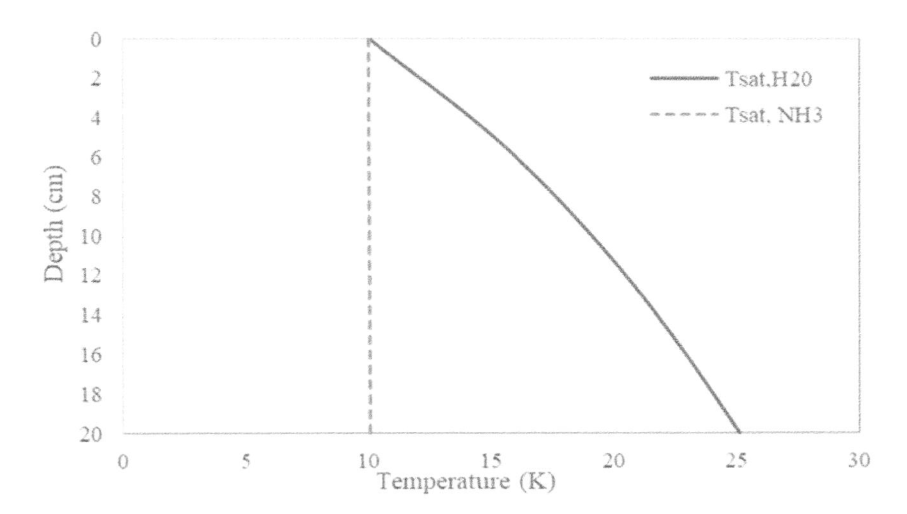

Figure 4.9. *Comparison of the impact of the weight of the water column (hydrostatic pressure) on the saturation pressure between water and ammonia for a fluid at equilibrium with vapor at 10°C. For a color version of this figure, see www.iste.co.uk/bonjour/refrigerators.zip*

$$\Delta T_{sat} = \frac{1{,}6\sigma T_W}{R_{cav}\rho_v \Delta h_{lv}} \quad [4.25]$$

Beyond the impact that the low vapor density has on the superheat necessary to trigger boiling and on the number of activatable sites, it also has an impact on the phenomena observed. Indeed, the vapor masses generated are in fact very large. Many authors have observed bubbles of several centimeters during pool boiling studies in a confined (Giraud et al. 2016) or not confined (Van Stralen et al. 1975; Michaïe et al. 2017) environment, whereas they are only a few millimeters at atmospheric pressure. The generation of these bulky masses, coupled with a generally subcooled boiling environment and the larger wall superheat required to trigger boiling, means that the waiting time between the generation of vapor masses may be several minutes, when it is usually several milliseconds (varies according to fluids and conditions). During these long waiting times, heat transfer takes place by conduction and convection. The heat transfer coefficients are therefore greatly degraded in comparison to those obtained during the growth of the vapor masses.

From a hydrodynamic point of view, with the volumetric flow rate of water vapor generated being high, the velocities of the vapor phases can be very high, leading to significant pressure drops. However, at these low pressures, a pressure drop of a few kilopascals results in an increase in the saturation temperature of several kelvins. Indeed, with the pressure being greater at the level of the free interface than at the evaporator outlet (the pressure drops decreasing the pressure along the exchanger in the direction of the flow), the pressure of corresponding saturation is therefore higher in the evaporator than at its outlet.

NOTE.– A pressure drop of 0.4 kPa–1.2 kPa leads to an increase in the saturation temperature of 5.9 K at the free interface.

To overcome the technical and scientific difficulties raised by the phase change at low pressures, the evaporators used in water systems are mainly falling film evaporators. These evaporators have the advantage of not creating a pool of water and not involving the boiling phenomenon. They exhibit high heat transfer coefficients, even at low superheats. However, these evaporators require a recirculation pump, an essential element for redistributing the fluid at the top of the evaporator, while guaranteeing a minimum flow rate to avoid drying out areas. This element adds a rotating part (the pump) to the entire installation, with all the associated issues (cost, maintenance, cavitation, additional electricity consumption, etc.). Thus, flooded evaporators with capillary films have emerged. These evaporators are used in both sorption heat pumps and desalination and sorption heat recovery processes. In these evaporators, the tubes are generally totally immersed at the start of the adsorption phase (Figure 4.10). Heat transfer takes place by evaporation of a film created and maintained on the tubes thanks to different surface structures, the structures acting as a pump by capillary effect in order to supply the film.

The heat transfer coefficient obtained in these exchangers depends on the immersion depth of the tubes. It is the lowest when the tube is totally submerged and at its maximum just before the bond between the film and the pool of water breaks ("relative filling level" < 0 in Figure 4.10). It increases by about a factor of 10 between these two instants. When the overall heat transfer coefficient is at its maximum, it can be up to twice as high as that obtained in falling film exchangers. Creating surfaces to obtain a high heat transfer coefficient while maintaining the film as long as possible on the surface is therefore currently one of the major challenges for these evaporators.

Despite their interesting performances, the evaporators used in sorption systems and presented above are generally of tubular geometry. However, these geometries have limitations as to the possible reduction of the cost and the size of these exchangers. Studies have therefore been carried out in order to promote plate heat exchangers under these operating conditions (low pressure), since plate heat exchangers usually have high compactness and low cost. The preliminary studies carried out (Giraud et al. 2016) show that the topology of two-phase flows is completely transformed at these low pressures. Although it is possible to initiate boiling phases, most of the thermal power exchanged is not due to the generation of vapor masses, but to the evaporation of a falling film created on the upper part of the evaporator by the explosion of the bubbles (Figure 4.11a). However, in addition to the technical issues related to the mechanical strength of the exchanger, the performance currently recorded leaves something to be desired: if the heat transfer coefficients before any surface treatment are comparable to those obtained in falling film towers, they display large cyclic variations (Figure 4.11b), and a huge part of the exchanger, essential to the generation of the bubbles and therefore to the creation of the film, has a low or even zero time-averaged energy density. Understanding the phenomena associated with low pressure boiling is therefore essential in order to eventually optimize this type of exchanger and obtain competitive and compact evaporators. This is the challenge of the studies currently being carried out on low-pressure boiling phenomena. Throughout this work, the researchers have aimed to both increase knowledge of the fundamental phenomena of the liquid/vapor phase change at low pressure and develop surface structures that make it possible to reduce the superheat required for the onset of boiling and increase the local heat transfer coefficient.

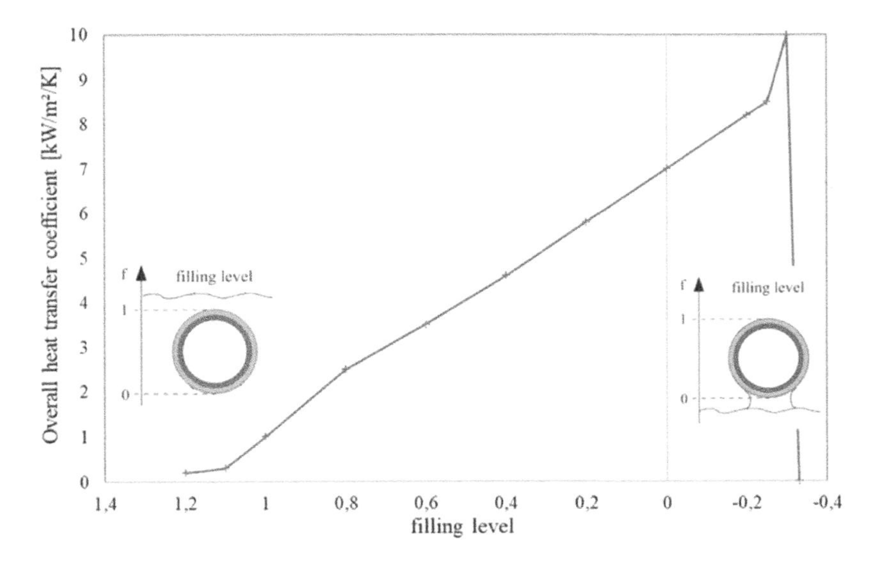

Figure 4.10. *Schematic evolution of the heat transfer coefficient obtained (secondary fluid temperature 15°C, pressure 1.2 kPa) (from Seiler et al. 2020). For a color version of this figure, see www.iste.co.uk/bonjour/refrigerators.zip*

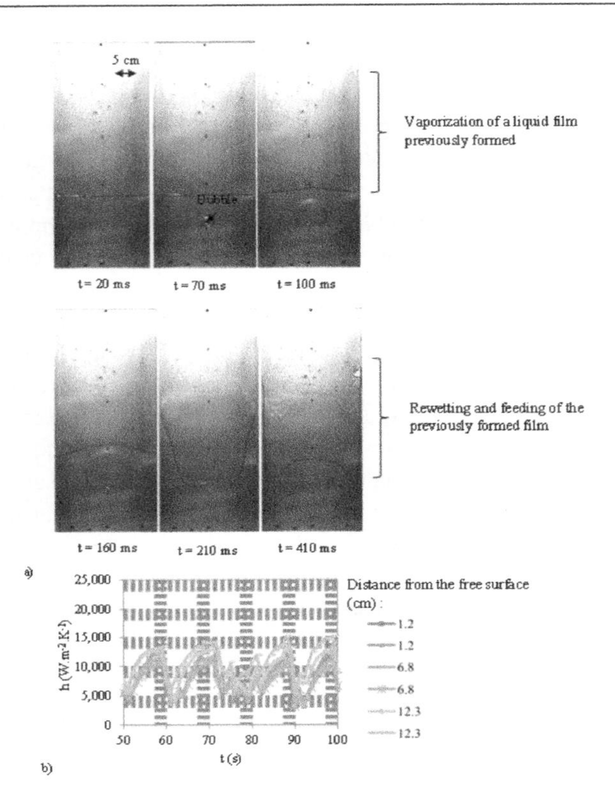

Figure 4.11. *Example a) of two-phase flow observed in a flat plate evaporator and b) of evolution of the heat transfer coefficient (secondary fluid temperature 21°C, pressure 1.4 kPa, Giraud et al. (2016)). For a color version of this figure, see www.iste.co.uk/bonjour/refrigerators.zip*

4.3.5. *Coupling of components: the evapo-absorber*

In absorption systems using water as refrigerant, for air conditioning applications, the low level of pressure within the evaporator and the absorber is a few millibars. It is therefore important to reduce the pressure drops related to the flow of water vapor between these two components. In this sense, it was planned to combine the two functions (evaporation of the refrigerant and absorption of it by the absorbent solution) in a single component: the evapo-absorber.

The evapo-absorber of absorption systems is a crucial component since the cooling capacity is directly related to the absorption rate.

The evapo-absorber of systems marketed for building applications generally consists of a large reservoir inside which the refrigerant is sprayed through a nozzle to promote evaporation, while the absorber solution flows nearby over a tube bundle.

To increase the compactness of the evapo-absorber, it is possible to envisage the construction of the evapo-absorber by an assembly of plates that water or the absorbent solution flow over, which is low in refrigerant (water/LiBr mixture). The plates are arranged so as to form an alternation of channels for the circulation of liquid refrigerant and for the absorbent solution. The vapor created is absorbed by the low-refrigerant solution, enriching it. To ensure the proper functioning of the absorption engine, a fluid recovery block is placed at the bottom of the plates. Its function is to recover each of the flows separately, then to combine the water and the water-enriched solution in two separate reservoirs. It should be noted that there is no direct contact of the liquid phases in the evapo-absorber.

Zinet et al. (2012) developed a numerical model of an engine using an evapo-absorber (Figure 4.12). This operation of the evapo-absorber is based on the mass and heat transfers of a falling film on the outer walls of solid plates.

The falling film model used was taken from the work by Auracher et al. (2008) and Flessner et al. (2009).

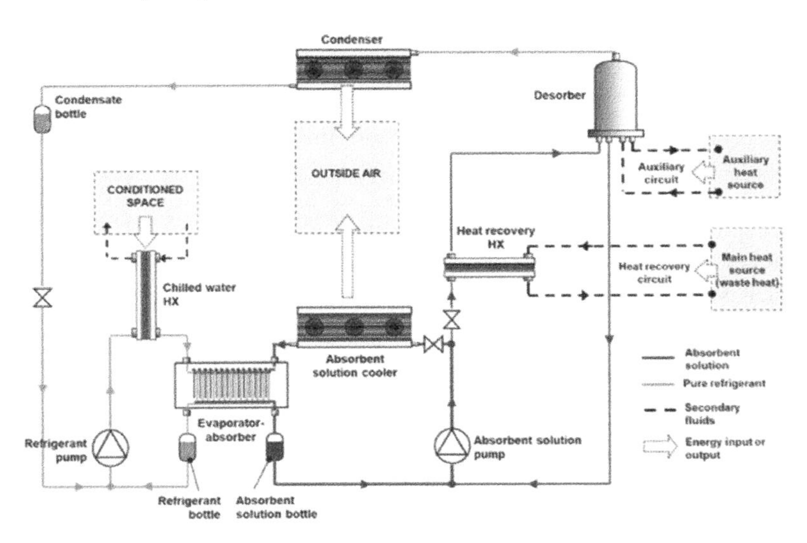

Figure 4.12. *Diagram of the absorption chiller modeled by Zinet et al. (2012)*

In this absorption cycle, only a small fraction of the refrigerant flow is evaporated and absorbed by the lithium bromide solution. The modeled principle of operation of the evapo-absorber is illustrated in Figure 4.13. It essentially consists of a set of parallel vertical plates inside an airtight enclosure. Each plate is fed by a horizontal distribution tube located on its upper edge. The wall of this tube is pierced in order to generate a film which flows over the entire surface of the plate.

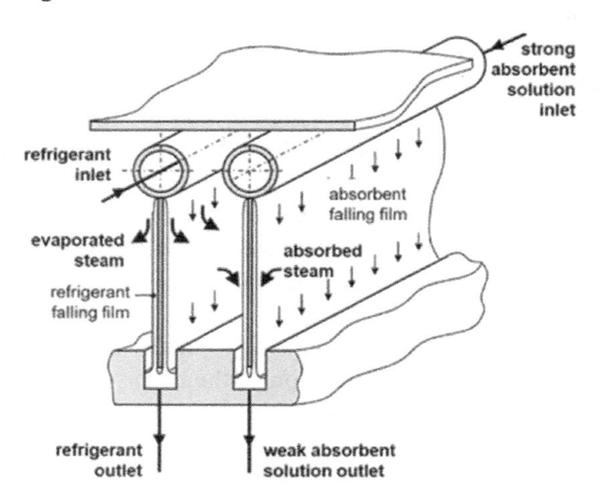

Figure 4.13. *Principle of operation of the evapo-absorber proposed by Zinet et al. (2012)*

Due to the difference between the pressure of the refrigerant at the inlet of the evapo-absorber, and the pressure inside the enclosure (lower than its saturation pressure), a fraction of the refrigerant vaporizes as soon as it enters the enclosure (flash evaporation). Another marginal fraction is then evaporated along the plate since saturated liquid falls downs, generating a slight pressure loss. The saturated refrigerant vapor accumulates inside the evapo-absorber and is absorbed on the surface of the film of the lithium bromide solution. As the system is globally adiabatic, the heat of absorption of water is collected in sensible form by the solution. To limit this heating, which reduces the absorption potential of the solution, a significant reflux (recirculation of the fluid) is imposed. Thus, the flow of solution circulating in the evapo-absorber is much higher than the flow that reaches the desorber.

The liquid water collected at the outlet of the evapo-absorber then circulates through a heat exchanger to provide the cooling effect. The temperature of the liquid water increases through the heat exchanger. This stream of water is mixed with the liquid water coming from the condenser before entering the evapo-absorber.

The lithium bromide solution enters the evapo-absorber, where the pressure is higher than its vapor pressure, causing absorption. The solution temperature and water concentration increase as absorption occurs. Therefore, the driving force for absorption decreases as the solution flows downward. Only a certain part of the solution flow then circulates to the desorber.

Boudard and Bruzzo (2010), Goulet (2011) and Obame Mve (2014) showed that the confinement of fluids in the evapo-absorber ensures the proper functioning of the system when it is subjected to vibrations or tilt. In this evapo-absorber, the two fluids circulate separately by gravity between two woven grids and are confined within the structure thanks to capillary forces (Figure 4.14). This confinement of the fluids makes it possible to avoid contact between the two liquid flows, which would induce significant losses in performance. This flow configuration has the advantage of offering large transfer areas in a relatively small volume.

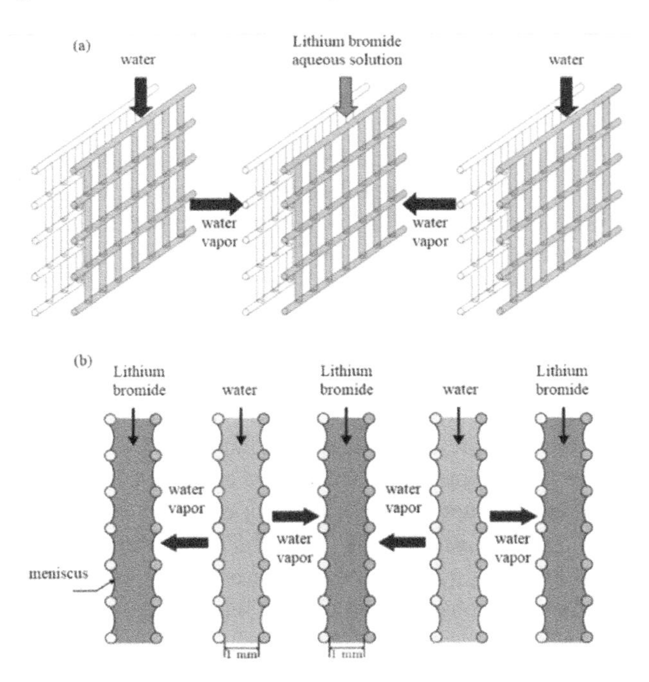

Figure 4.14. *Description of how the evapo-absorber works: (a) Perspective view of three plates, (b) 2D view of a 5-plate prototype. (Obame Mve 2014). For a color version of this figure, see www.iste.co.uk/bonjour/refrigerators.zip*

The water vapor, resulting from the partial evaporation of water at low pressure, is absorbed by the saline solution of lithium bromide at the liquid–vapor interface formed by the menisci. These menisci, which form the transfer surface, result from a balance between the capillary forces, the forces of inertia, the viscous forces and the forces of interaction with the grid.

An improvement of mass and heat transfers and, consequently, an increase in the absorption rate can be obtained by the occurrence of instabilities in the flow (turbulence) and by the increase in the surface area of the menisci, which constitutes the absorption surface, and which strongly depends on the geometry of the confinement grids and on the wettability.

The reduction in the absorption driving force is due to the heat released by the exothermic vapor absorption reaction. This observation demonstrates the interest of cooling the solution during the absorption process (circulation of a secondary fluid), so as to extract the heat generated and, therefore, increase the absorption rate.

4.4. Risks associated with liquid–solid phase transition phenomena

4.4.1. *Crystallization*

Absorption processes are traditionally designed so that the absorbent solution remains only in the liquid phase. One of the limitations of the use of saline solutions concerns the crystallization of the salt when the concentration exceeds saturation conditions. The threshold for the occurrence of nucleation must be estimated in order to avoid the appearance of these saturated states.

The Oldham diagram, which represents the saturation conditions (pressure, temperature, composition) of the liquid solution, is the most used and the most practical for studying the absorption cycle. It gives the solute fraction of the solution as a function of temperature and pressure. Figure 4.15 represents the Oldham diagram of the LiBr–H_2O couple. The 0% line of constant composition corresponds to the liquid–vapor equilibrium of pure water. For example, if the temperature of the solution can reach a minimum value of 10°C, the mass fraction must be limited to 0.58 kg LiBr/kg solution to guarantee the absence of crystals in the solution: this mass fraction is indeed the solubility limit of the solution at this temperature.

The crystallization phenomenon can therefore, for example, appear when the saline solution entering the absorber has a salt concentration or a temperature that is too high.

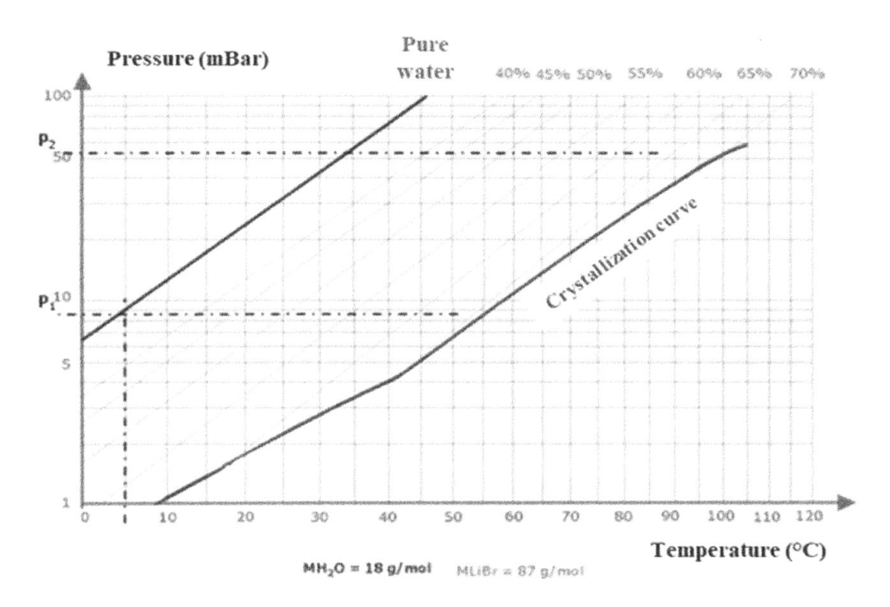

Figure 4.15. *Oldham diagram of the water/lithium bromide pair. For a color version of this figure, see www.iste.co.uk/bonjour/refrigerators.zip*

NOTE.– As soon as the crystallization phenomenon appears in the cycle, it is imperative to implement procedures to dissolve the crystals in order to avoid any risk of obstruction in the exchangers.

Different procedures to address the crystallization problem in LiBr/H_2O absorption systems have been described by Wang et al. (2011). One of the most effective identified technologies is the use of crystallization inhibitors. The most widely used chemical inhibitor today to reduce the risk of crystallization from lithium bromide solution is ethylene glycol. The commercial water–ethylene glycol–lithium bromide mixture, evaluated by Rivera et al. (1999), proved to be a potential replacement for the H_2O–LiBr pair, since it allowed similar performances to be obtained.

NOTE.– The absorption phenomenon can also be taken advantage of in thermal storage processes. In this case, the crystallization phenomenon can become an asset. At any given time, only a minority fraction of the solution is circulating in the system. The fraction that does not circulate is then stored in a dedicated reservoir. N'Tsoukpoe et al. (2014) considered an innovative operation allowing the increase in the specific energy of the storage, by accepting a crystallization of part of the solution in this reservoir during the cycle, if these crystals can then be dissolved.

4.4.2. *Freezing*

When water is used as the working fluid in sorption-based reverse cycle engines, freezing phenomena may occur. This freezing can be observed when the operating conditions cause one of the components (apart from the components in which there is an absorbent/working fluid or adsorbent/working fluid pair) to operate below the triple point of water. This is the case, for example, when the adsorber is oversized (it will then tend to impose its operating pressure) or when the cold source temperature is below 0°C. These constraints therefore require appropriate sizing of the various components (including circulation lines) and limit the use of these systems to applications for which the temperature within the condenser and the evaporator are positive (> 0°C). It also imposes, as a preventive measure, the elimination of pairs using water as a working fluid for reverse cycle engines used as heat pumps. Thus, the development of sorption systems using water as a refrigerant for applications such as heat pumps or refrigeration is greatly hampered.

To overcome these constraints, different solutions have been studied. The most common is to add a second component to the water to lower the triple point temperature of the mixture. There are commercial installations where this principle has been used successfully to evaporate the refrigerant at temperatures down to -15°C.

In most of these installations, the working fluid is an aqueous solution of lithium bromide with a mass fraction of lithium bromide of between 4 and 30%, depending on the applications. Other substances can be used, such as alcohols (ethylene glycol, propylene glycol, ethanol), ethers (1,4-dioxane) or salts ($CaCl_2$, NaCl, LiCl).

In addition to compatibility issues between the mixture and the installation (as is the case, for example, when using an aqueous solution of 1,4-dioxane – a very corrosive mixture), a deterioration in the performance of the installation with respect to the use of pure water is generally observed for positive temperatures. Several phenomena can explain this degradation:

– the existence of an additional resistance to heat transfer due to the migration of species in the fluid (mass diffusion);

– the modification of the operating pressure of the evaporator reducing the ab-adsorption capacity of the system (decrease in the sorption loading estimated by

Seiler et al. (2017) at approximately 6% when using an aqueous solution containing 60% (by mass) of glycol at 5°C);

– the coab(ad)sorption of the two species modifying the sorption dynamics and the equilibrium of the system;

– the modification of the thermophysical properties, resulting in a modification of the phenomena observed (reduction in the size of the bubbles, disappearance of the unstable phenomena observed during the vaporization of water at low pressure, reduction in the wettability of the exchange surfaces in the case of falling film formation). The variation in the thermophysical properties by the addition of an alcoholic additive may, however, be desired. For example, Lonardi and Luke (2017) show opportunities for improving mass and heat transfer in ab and adsorbers (see section 4.2.1.2).

Changes in the composition of the mixture over time can also be observed, leading in certain cases to a gradual freezing of the installation, then to its complete shutdown after several operating hours (Odashima et al. 2017). These phenomena are currently being studied to enable long-term operation of the installations.

4.5. Conclusion

The design of reverse cycle thermal compression engines, i.e. chillers or heat pumps, is based on a good understanding of various complex physical phenomena. In addition to liquid-vapor phase change heat transfers present in the condenser and evaporator, coupled heat and mass transfers govern the operation of sorption reactors (absorber, adsorber, desorber). The current technologies of these reactors are numerous and this chapter highlights the particular constraints that their use generates. The development of these systems will require reliable correlations to be obtained, making it possible to predict the behavior of the reactors through the development and optimization of innovative designs specific to the targeted applications.

To promote the integration of sorption-based reverse cycle engines, it is important to quantify the overall performance of this type of system, taking into account the quality of the energies involved in order to thermodynamically justify their performance. The development of these systems will need to demonstrate not only their economic relevance, but also their adaptation to alternative energies (use and recovery of renewable or fatal energy), which must now take a greater part in the global energy mix.

4.6. References

Altamirano, A., Le Pierrès, N., Stutz, B. (2019). Review of small-capacity single-stage continuous absorption systems operating on binary working fluids for cooling: Theoretical, experimental and commercial cycles. *International Journal of Refrigeration*, 106, 350–373.

Aristov, Y.I., Glaznev, I.S., Girnik, I.S. (2012). Optimization of adsorption dynamics in adsorptive chillers: Loose grains configuration. *Energy*, 46, 484–492. doi: 10.1016/j.energy.2012.08.001.

Arzoz, D., Rodriguez, P., Izquierdo, M. (2005). Experimental study on the adiabatic absorption of water vapor into LiBr/H_2O solutions. *Applied Thermal Engineering*, 25, 797–811.

Auracher, H., Wohlfeil, A., Ziegler, F. (2008). A simple model for steam absorption into a falling film of aqueous lithium bromide solution on a horizontal tube. *Heat and Mass Transfer*, 44, 1529–1536.

Bankoff, S.G. (1971). Minimum thickness of a draining liquid film. *International Journal of Heat and Mass Transfer*, 14, 2143–2146.

Bendix, P., Füldner, G., Möllers, M., Kummer, H., Schnabel, L., Henninger, S., Henning, H. (2017). Optimization of power density and metal-to-adsorbent weight ratio in coated adsorbers for adsorptive heat transformation applications. *Applied Thermal Engineering*, 124, 83–90. doi: 10.1016/j.applthermaleng.2017.05.165.

Bird, R.B., Stewart, W.E., Lightfoot, E.N. (2007). *Transport Phenomena*. John Wiley & Sons, New York.

Boudard, E. and Bruzzo, V. (2010). Dispositif de climatisation par absorption perfectionné. Patent, UE 2 871 221. PSA Peugeot Citroen.

Brauner, N. (1991). Non isothermal vapour absorption into falling film. *International Journal of Heat and Mass Transfer*, 34, 767–784.

Calabrese, L., Bonaccorsi, L., Bruzzaniti, P., Proverbio, E., Freni, A. (2019). SAPO-34 based zeolite coatings for adsorption heat pumps. *Energy*, 187, 115981.

Carey, V.P. (1989). *Liquid-Vapor Phase Change Phenomena – An Introduction to the Thermophysics of Vaporization and Condensation Processes in Heat Transfer Equipment*. Taylor and Francis, Boca Raton, FL.

Chengming, S., Qinghua, C., Tien-Chien, J., Wang, Y. (2010). Heat transfer performance of lithium bromide solution in falling film generator. *International Journal of Heat and Mass Transfer*, 53, 3372–3376.

Dawoud, B. (2007). On the effect of grain size on the kinetics of water vapour adsorption and desorption into/from loose pellets of FAM-Z02 under a typical operating condition of adsorption heat pumps. *Journal of Chemical Engineering of Japan*, 40(13), 1298–1306. Presented at the *International Symposium on Innovative Materials for Processes*.

Dawoud, B. (2013). Water vapor adsorption kinetics on small and full scale zeolite coated adsorbers; A comparison. *Applied Thermal Engineering*, 50(2), 1645–1651.

Deng, S.M. and Ma, W.B. (1999). Experimental studies on the characteristics of an absorber using LiBr/H2O solution as working fluid. *International Journal of Refrigeration*, 22, 293–301.

Duong, X.Q., Cao, N.V., Lee, W.S., Park, M.Y., Chung, J.D., Bae, K.J., Kwon, O.K. (2020). Effect of coating thickness, binder and cycle time in adsorption cooling applications. *Applied Thermal Engineering*, 116265.

Flessner, C., Petersen, S., Ziegler, F. (2009). Simulation of an absorption chiller based on a physical model. *7th Modelica Conference*, Como.

Florides, G.A., Kalogirou, S.A., Tassou, S.A., Wrobel, L.C. (2003). Design and construction of a LiBr-water absorption machine. *Energy Conversion and Management*, 44(15), 2483–2508.

Freni, A., Sapienza, A., Glaznev, I.S., Aristov, Y.I., Restuccia, G. (2012). Experimental testing of a lab-scale adsorption chiller using a novel selective water sorbent "silica modified by calcium nitrate". *International Journal of Refrigeration*, 35(3), 518–524. doi: 10.1016/j.ijrefrig.2010.05.015.

Giraud, F., Toublanc, C., Rullière, R., Bonjour, J., Clausse, M. (2016). Experimental study of water vaporization occurring inside a channel of a smooth plate-type heat exchanger at subatmospheric pressure. *Applied Thermal Engineering*, 106, 180–191.

Glaznev, I.S. and Aristov, Y.I. (2010). The effect of cycle boundary conditions and adsorbent grain size on the water sorption dynamics in adsorption chillers. *International Journal of Heat and Mass Transfer*, 53(9–10), 1893–1898.

Golparvar, B., Niazmand, H., Sharafian, A., Hosseini, A.A. (2018). Optimum fin spacing of finned tube adsorber bed heat exchangers in an exhaust gas-driven adsorption cooling system. *Applied Energy*, 232, 504–516.

Goulet, R. (2011). Development and analysis of an innovative evaporator/absorber for automotive absorption-based air conditioning systems: Investigation on the simultaneous heat and mass transfer. PhD Thesis, INSA Lyon.

Grossman, G. (1983). Simultaneous heat and mass transfer in film absorption under laminar flow. *International Journal of Heat and Mass Transfer*, 26, 357–371.

Grossman, G. (1986). *Handbook of Heat and Mass Transfer*, 2nd edition. Gulf Publishing Co., Houston, TX.

Grossman, G. (1987). Analysis of interdiffusion in film absorption. *International Journal of Heat and Mass Transfer*, 30, 205–208.

Herold, K.E., Radermacher, R., Klein, S.A. (1996). *Absorption Chillers and Heat Pumps*. CRC Press, Boca Raton, FL.

Hihara, E. and Saito, T. (1992). Effect of surfactant on falling film absorption. *International Journal of Refrigeration*, 16, 339–346.

Hozawa, M., Inoue, M., Sato, J., Tsukada, T., Imaishi, N. (1991). Marangoni convection during steam absorption into aqueous LiBr solution with surfactant. *Journal of Chemical Engineering of Japan*, 24, 209–214.

Hsu, Y.Y. (1962). On the size range of active nucleation cavities on a heating surface. *Journal of Heat Transfer*, 84, 207–213.

Hu, X. and Jacobi, A.M. (1996). The intertube falling film: Part 1 – Flow characteristics mode transitions and hysteresis. *Journal of Heat Transfer*, 118, 616–633.

Ingle, S.E. and Horne, F.H. (1973). The Dufour effect. *The Journal of Chemical Physics*, 59, 5882–5894.

Jeong, S. and Garimella, S. (2002). Falling-film and droplet mode heat and mass transfer in a horizontal tube LiBr/water absorber. *International Journal of Heat and Mass Transfer*, 45, 1445–1458.

Kang, Y.T., Kim, H.J., Lee, K.I. (2008). Heat and mass transfer enhancement of binary nanofluids for H2O/LiBr falling film absorption process. *International Journal of Refrigeration*, 31, 850–856.

Kashiwagi, T., Kurosaki, Y., Shishido, H. (1985). Enhancement of vapour absorption into a solution using the Marangoni effect. *Transactions of the JSME*, 51(463), 1002–1009.

Kawae, N., Shigechi, T., Kanemaru, K., Yamada, T. (1989). Water vapor evaporation into laminar film flow of a lithium bromide-water solution (influence of variable properties and inlet film thickness on absorption mass transfer rate). *Heat Transfer – Japanese Research*, 18, 58–70.

Keizer, C. (1982). Absorption refrigeration machines. PhD Thesis, Delft University of Technology.

Killion, J.D. and Garimella, S. (2001). A critical review of models of coupled heat and mass transfer in falling-film absorption. *International Journal of Refrigeration*, 24, 755–797.

Kim, D.S. and Infante Ferreira, C.A. (2008). Flow patterns and heat and mass transfer coefficients of low Reynolds number falling film on vertical plates: Effects of wire screen and an additive. *International Journal of Refrigeration*, 32, 138–149.

Kim, K.J., Berman, N.S., Chau, D.S.C., Wood, B.D. (1995). Absorption of water vapour into falling films of aqueous lithium bromide. *International Journal of Refrigeration*, 18(7), 486–494.

Kim, K.J., Berman, N.S., Wood, B.D. (1996). The interfacial turbulence in falling film absorption: Effects of additives. *International Journal of Refrigeration*, 19, 322–330.

Kyung, I., Herold, K.E., Kang, Y.T. (2007). Experimental verification of H2O/LiBr absorber bundle performance with smooth horizontal tubes. *International Journal of Refrigeration*, 30, 582–590.

Lonardi, F. and Luke, A. (2017). Adsorption mechanisms of alcoholic additives in water at low pressure: An experimental study. *International Sorption Heat Pump Conference*, ISHPC2017, Tokyo, 1053.

Meunier, F. (2016). Systèmes thermiques à sorption solide. *Techniques de l'Ingénieur*, BE9737, 1, 1–19.

Michaïe, S., Rullière, R., Bonjour, J. (2017). Experimental study of bubble dynamics of isolated bubbles in water pool boiling at subatmospheric pressures. *Experimental Thermal and Fluid Science*, 87, 117–128.

Mitra, S., Muttakin, M., Thu, K., Saha, B.B. (2018). Study on the influence of adsorbent particle size and heat exchanger aspect ratio on dynamic adsorption characteristics. *Applied Thermal Engineering*, 133, 764–773.

Miyazaki, T., Akisawa, A., Saha, B.B., El-Sharkawy, I.I., Chakraborty, A (2009). A new cycle time allocation for enhancing the performance of two-bed adsorption chillers. *International Journal of Refrigeration*. 32(5), 846–853.

Nakoryakov, V.E. and Grigor'eva, N.I. (1977). Combined heat and mass transfer during absorption in drops and films. *Journal of Engineering Physics*, 32, 243–247.

Nakoryakov, V.E., Bufetov, N.S., Grigorieva, N.I., Dekhtyar, R.A., Marchuk, I.V. (2004). Vapor absorption by immobile solution layer. *International Journal of Heat and Mass Transfer*, 47, 1525–1533.

Nakoryakov, V.E., Grigoryeva, N.I., Bufetov, N.S., Dekhtyar, R.A. (2008). Heat and mass transfer intensification at steam absorption by surfactant additives. *International Journal of Heat and Mass Transfer*, 51, 5175–5181.

Niazmand, H. and Dabzadeh, I. (2012). Numerical simulation of heat and mass transfer in adsorbent beds with annular fins. *International Journal of Refrigeration*, 35(3), 581–593.

Niazmand, H., Talebian, H., Mahdavikhah, M. (2013). Effects of particle diameter on performance improvement of adsorption systems. *Applied Thermal Engineering*, 59(1–2), 243–252.

N'Tsoukpoe, K.E., Perier-Muzet, M., Le Pierrès N., Luo, L., Mangin, D. (2014). Thermodynamic study of a LiBr–H2O absorption process for solar heat storage with crystallisation of the solution. *Solar Energy*, 104, 2–15.

Obame Mve, H. (2014). Compréhension des écoulements et optimisation des transferts de chaleur et de masse au sein d'une structure capillaire. PhD Thesis, INSA Lyon.

Odashima, S., Noda, H., Takahashi, S. (2017). Generation of the below zero degree temperature by use of the absorption refrigerator with the water-1,4-dioxane refrigerant. *International Sorption Heat Pump Conference*, ISHPC2017, Tokyo, 1040.

Patnaik, V. and Perez-Blanco, H. (1996). Roll waves in falling films: An approximate treatment of the velocity field. *International Journal of Heat and Fluid Flow*, 17, 63–70.

Platten J.K. (2006). The Soret effect: A review of recent experimental results. *Journal of Applied Mechanics*, 73, 5–15.

Restuccia, G., Freni, A., Vasta, S., Aristov, Y. (2004). Selective water sorbent for solid sorption chiller: Experimental results and modelling. *International Journal of Refrigeration*, 27(3), 284–293.

Rivera, W., Romero, R., Best, R., Heard, C. (1999). Experimental evaluation of a single-stage heat transformer operating with the water/Carrol mixture. *Energy*, 24(4), 317–323.

Rivera, W., Cardoso, M., Romero, R. (2001). Single-stage and advanced absorption heat transformers operating with lithium bromide mixtures used to increase solar pond's temperature. *Solar Energy Materials and Solar Cells*, 70(3), 321–333.

Santamaria, S., Sapienza, A., Frazzica, A., Freni, A., Girnik, I.S., Aristov, Y.I. (2014). Water adsorption dynamics on representative pieces of real adsorbers for adsorptive chillers. *Applied Energy*, 134, 11–19.

Schnabel, L., Füldner, G., Velte, A., Laurenz, E., Bendix, P., Kummer, H., Wittstadt, U. (2018). Innovative adsorbent heat exchangers: Design and evaluation. In *Innovative Heat Exchangers*, Bart, H.J. and Scholl, S. (eds). Springer, Cham. doi: 10.1007/978-3-319-71641-1_12.

Seiler, J., Hackmann, J., Lanzerath, F., Bardow, A. (2017). Refrigeration below zero C: Adsorption chillers using water with ethylene glycol as antifreeze. *International Journal of Refrigeration*, 77, 39–47.

Seiler, J., Volmer, R., Krakau, D., Poehls, J., Ossenkopp, F., Schnabel, L., Bardow, A. (2020). Capillary-assisted evaporation of water from finned tubes – Impacts of experimental setups and dynamics. *Applied Thermal Engineering*, 165, 114620.

Sharafian, A. and Bahrami, M. (2015). Critical analysis of thermodynamic cycle modeling of adsorption cooling systems for light-duty vehicle air conditioning applications. *Renewable and Sustainable Energy Reviews*, 48, 857–869.

Shoushi, B., Xuehu, M., Zhong, L., Jiabin, C., Hongxia, C. (2010). Numerical simulation on the falling film absorption process in a counter-ow absorber. *Chemical Engineering Journal*, 156, 607–612.

Takamatsu, H., Yamashiro, H., Takata, N., Honda, H. (2003). Vapor absorption by LiBr aqueous solution in vertical smooth tubes. *International Journal of Refrigeration*, 26, 659–666.

Van Stralen, S.J.D., Cole, R., Sluyter, W.M., Sohal, M.S. (1975). Bubble growth rates in nucleate boiling of water at subatmospheric pressures. *International Journal of Heat and Mass Transfer*, 18, 655–669.

Wang, K., Abdelaziz, O., Kisari, P., Vineyard, E.A. (2011). State-of-the-art review on crystallization control technologies for water/LiBr absorption heat pumps. *International Journal of Refrigeration*, 34(6), 1325–1337.

Wang, Y., Rullière, R., Revellin, R., Haberschill, P. (2015). A review of the experimental performances and challenges of the absorption system technologies. *24th International Congress of Refrigeration*, Yokohama.

Wu, W.T., Yang, Y.M., Maa, J.R. (1998). Effect of surfactant additive on pool boiling of concentrated lithium bromide solution. *International Communications in Heat and Mass Transfer*, 25(8), 1127–1134.

Yao, W., Bjurstroem, H., Setterwall, F. (1991). Surface tension of lithium bromide solutions with heat-transfer additives. *Journal of Chemical and Engineering Data*, 36, 96–98.

Zajaczkowski, B. (2016). Optimizing performance of a three-bed adsorption chiller using new cycle time allocation and mass recovery. *Applied Thermal Engineering*, 100, 744–752.

Ziegler, F. and Grossman, G. (1996). Heat transfer enhancement by additives. *International Journal of Refrigeration*, 5, 301–309.

Zinet, M., Rullière, R., Haberschill, P. (2012). A numerical model for the dynamic simulation of a recirculation single-effect absorption chiller. *Energy Conversion and Management*, 62, 51–63.

5

Magnetocaloric Refrigeration: Principle and Applications

Monica SIROUX

ICUBE UMR 7357, INSA Strasbourg, CNRS, Université de Strasbourg, France

5.1. Introduction

Magnetocaloric refrigeration, also known as magnetic refrigeration, is a technology that is considered a possible future alternative to our current refrigeration systems. This is why magnetic refrigeration has been the subject of research and development for many years. It is an emerging cooling technology that offers many advantages: energy efficiency greater than that of a conventional thermodynamic cycle, absence of atmospheric pollutants, absence of noise and compactness. This technology is based on the magnetocaloric effect (MCE), which results in the cooling of certain materials under the action of a magnetic field. To do this, we use materials such as gadolinium and alloys, characterized by a "giant" MCE. The MCE is maximum when the temperature of the material is close to its Curie temperature. Current research on magnetic refrigeration focuses on magnetocaloric materials, regenerators (the main element of a magnetic refrigerator) and magnets. Recently, many demonstrators and prototypes have been developed.

Refrigerators, Heat Pumps and Reverse Cycle Engines,
coordinated by Jocelyn BONJOUR.

This chapter presents an overview of magnetic refrigeration. The first part of this chapter deals with the theory of magnetic refrigeration (its history, MCE, magnetocaloric cycles, magnetocaloric materials). Then, a state of the art of the models is presented. Finally, the main realized magnetocaloric prototypes are presented, as well as future applications.

5.2. Magnetic refrigeration

5.2.1. *Overview*

The MCE in iron was discovered by E. G. Warburg in 1881 (Warburg 1881). The use of the MCE for energy conversion was then the subject of two patents: a magneto-thermal generator (Edison 1892) and a magneto-thermal motor (Tesla 1889). In 1889, N. Tesla's idea involves subjecting a magnetic body to the action of heat and bringing it to a temperature such that a mechanical action can set it in motion and thus allow electrical generation.

In 1949, the Nobel Prize in Chemistry was awarded to F. Giauque for his work in the field of magnetic refrigeration. The first room-temperature magnetic refrigerator was designed by Brown in 1976 (Brown 1976). The system developed by Brown consisted of parallel gadolinium plates between which circulated a heat transfer fluid (water and alcohol). With this device, Brown obtained a temperature difference of 47 K, using a superconducting magnet to generate a magnetic field of 7 T (Brown 1976).

In 1978, Stevert (1978) developed the concept of active regeneration, which consists of using the magnetocaloric material as a regenerator to amplify the temperature difference at each cycle.

In 1997, the giant magnetocaloric effect was discovered by Pecharsky and Gschneidner (1997) in a gadolinium–germanium–silicon alloy. From this date on, scientists and refrigeration industrialists considered magnetic refrigeration at room temperature to be an industrial application.

In 1988, Zimm (1988) produced the first magnetic refrigeration prototype at room temperature, characterized by a power compatible with industrial applications. After this, several prototypes were made around the world.

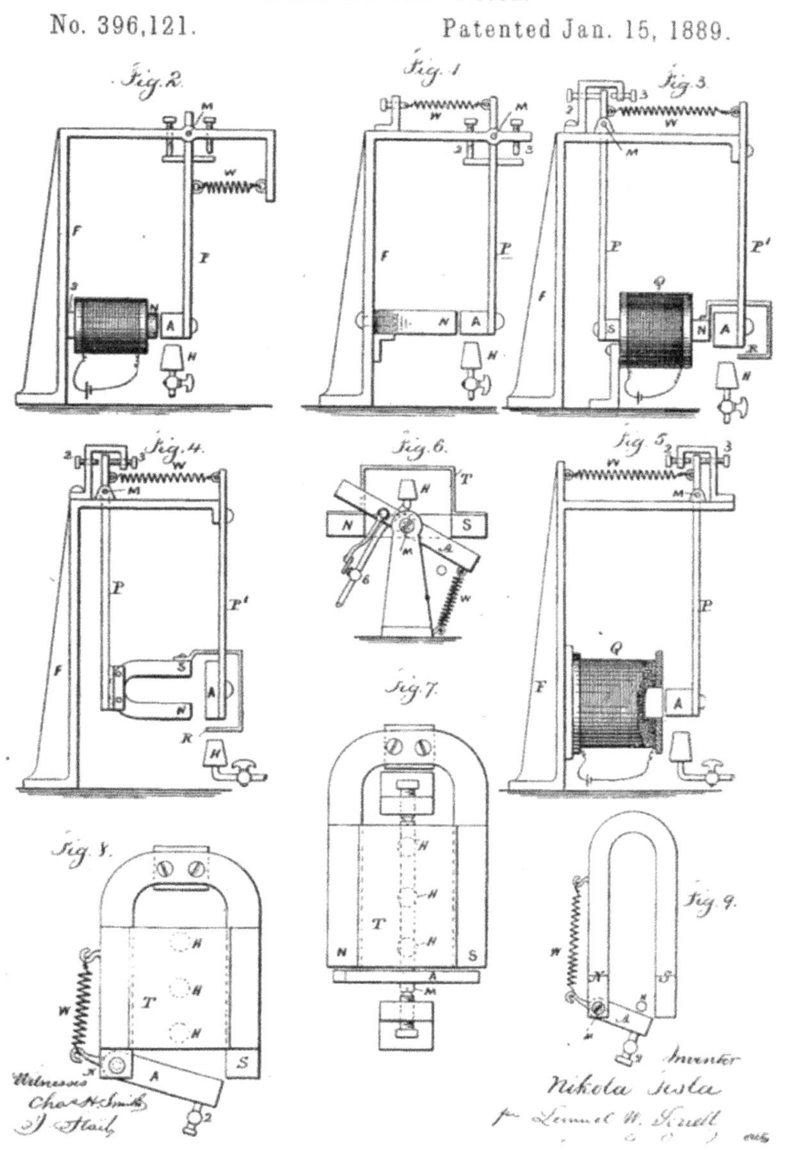

Figure 5.1. *Tesla's thermomagnetic motor (Tesla 1889)*

5.2.2. *The magnetocaloric effect*

Magnetic refrigeration has been used for decades for cryogenic applications. However, recent developments in materials science have made this technology a candidate for refrigeration at room temperature.

The MCE is an intrinsic property of magnetic materials that results in an instantaneous and reversible variation in their temperature or entropy when they are subjected to magnetic field variations. This effect is maximum at the transition temperature from the ferromagnetic phase (ordered state) to the paramagnetic phase (disordered state), known as the Curie temperature (or point) or the Néel temperature.

Figure 5.2 presents the MCE. If a magnetic material is exposed to a magnetic field, then the magnetic spins of the electrons in the atoms, initially disoriented, align. This reaction is exothermic and heat can be released to the environment. When the magnetic field is then removed, this process becomes endothermic and the temperature of the material will decrease because the magnetic spins become disoriented. The heat from the thermal load can be extracted using a heat transfer medium such as water, air or other substances depending on the application (Bouchekara 2008).

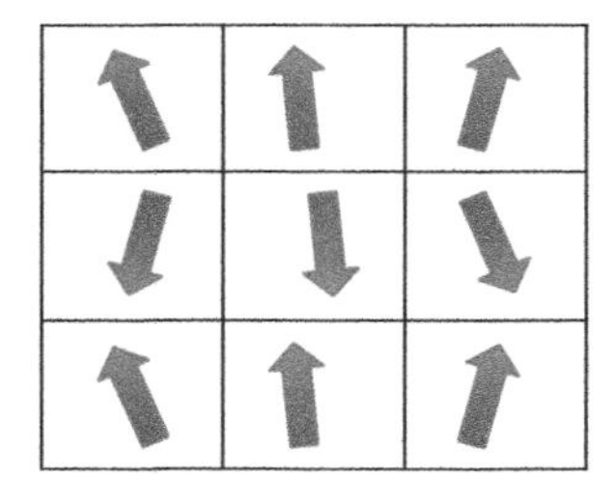
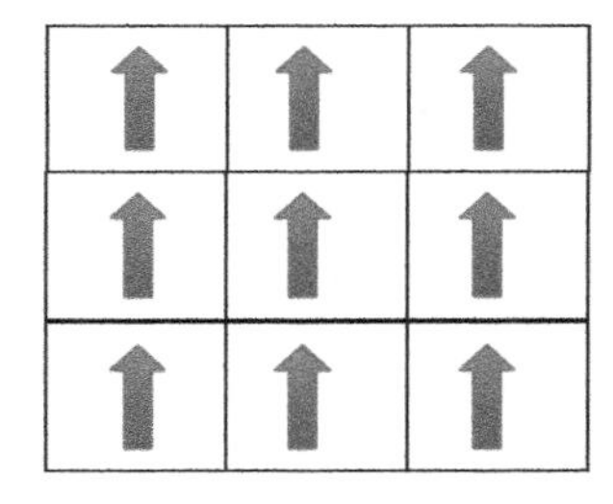

Figure 5.2. *Principle of the MCE. Left: magnetic material, the magnetic moments are in random directions. Right: magnetic material under magnetic field, the magnetic moments are aligned*

The total entropy S can be considered as the sum of three quantities: the magnetic entropy *Sm*, the network entropy *Sr* and the electronic entropy *Se*. The MCE can be quantified as the reversible change in material temperature (ΔT_{ad}) under the influence of a varying magnetic field in an adiabatic process, or the reversible change of magnetic entropy (ΔS_m) in an isothermal process:

$$S = Sm + Sr + Se \qquad [5.1]$$

The thermodynamic potential G of a material subjected to a magnetic field H can be written as:

$$G(P, H, T) = U + PV + \mu_0 HM - TS \qquad [5.2]$$

with:

– H, the magnetic field ($A \cdot m^{-1}$);

– M, the magnetization ($A \cdot m^{-1}$);

– P, the pressure (Pa);

– U, the internal energy (J);

– V, the volume (m^3);

– μ_0, the magnetic permeability of vacuum ($kg \cdot m \cdot A^{-2} \cdot s^{-2}$).

The expression of the variation in the internal energy of the material is given by:

$$dU(P, H, T) = TdS + \mu_0 HdM - PdV \qquad [5.3]$$

The following is then obtained:

$$dG(P, H, T) = VdP - \mu_0 HdM - SdT \qquad [5.4]$$

Schwarz's theorem for an exact total differential can be written as:

$$\left(\frac{\partial S}{\partial H}\right)_T = \mu_O \left(\frac{\partial M}{\partial T}\right)_H \qquad [5.5]$$

When the field varies from H_1 to H_2, the isothermal integration gives:

$$\Delta S_m(T, H_1, H_2) = \mu_O \int_{H1}^{H2} \left(\frac{\partial M}{\partial T}\right)_H dH \qquad [5.6]$$

For a constant temperature T, if the magnetic field varies from $\Delta H = H_2\text{-}H_1$, then the variation in isothermal magnetic entropy *ΔSm (T, ΔH)* is a function of the derivative of the magnetization with respect to the temperature.

The value $\left(\frac{\partial M}{\partial T}\right)$ is maximal around the Curie temperature *C*.

5.2.3. *Magneto-thermodynamic cycles*

The magnetic refrigerator is a reverse thermal engine (Figure 5.3) that operates according to a magneto-thermodynamic cycle.

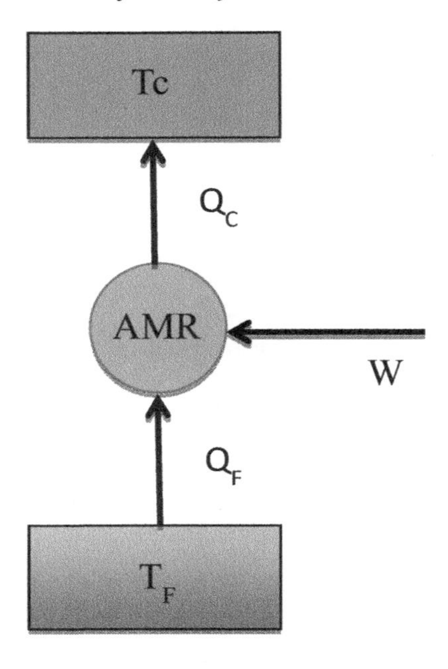

Figure 5.3. *Principle of a magnetic refrigerator*

A comparison between a magnetic cycle and a conventional refrigeration cycle is shown in Figure 5.4. The thermodynamic cycle involving the compression/expansion of the refrigerant gas used in conventional refrigeration systems is replaced by a thermomagnetic cycle with magnetization/demagnetization of a material with an MCE, which acts as a refrigerant. Solid magnetization/demagnetization processes are associated with heat absorption/rejection for heat exchangers. The work required to drive the magnetic refrigerator is associated with the actuator and the pump.

It is important to specify that, unlike conventional refrigeration, the processes of magnetization (compression) and demagnetization (relaxation) are almost reversible. This reversibility makes magnetic refrigeration a potentially more efficient process than gas compression and expansion. The efficiency of magnetic refrigeration can therefore be higher than that of conventional refrigerators.

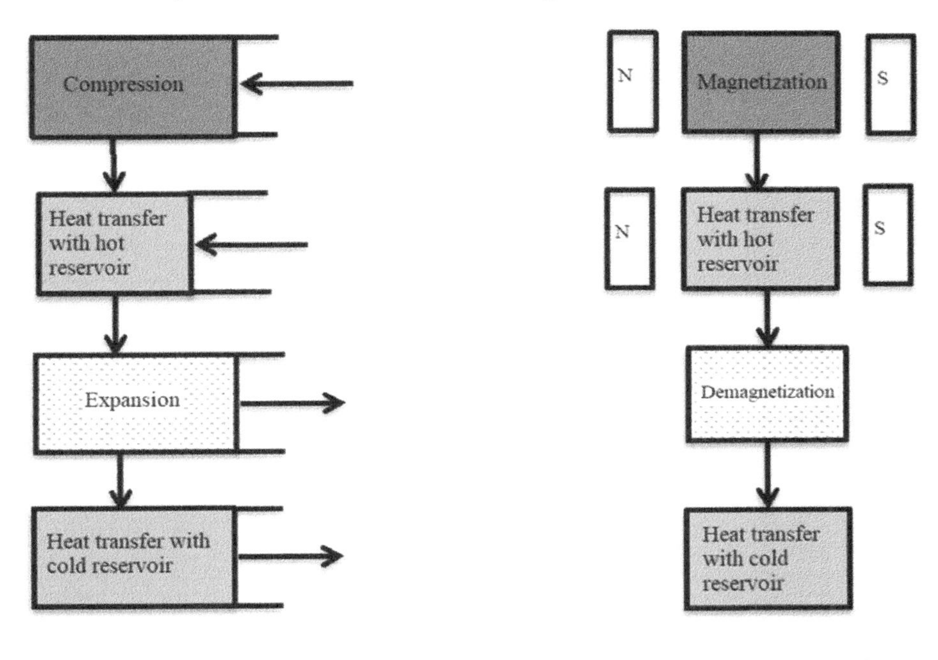

Figure 5.4. *Comparison between magnetic refrigeration and conventional refrigeration*

The conservation of energy for a magneto-thermodynamic cycle can be written as:

$$Q_C + Q_F + W = 0 \quad [5.7]$$

The coefficient of performance (COP) is given by:

$$COP = Q_F / W \quad [5.8]$$

with:

– Q_C, the heat transferred with the hot reservoir (J);

– Q_F, the heat transferred with the cold reservoir (J);

– W, the input work (J).

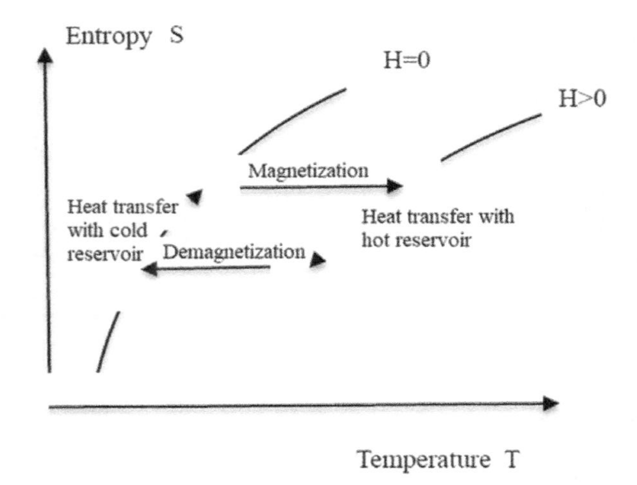

Figure 5.5. *Magnetic refrigeration cycle*

Different magnetic refrigeration cycles can be performed:

– the Carnot cycle;

– the Ericsson magnetic cycle;

– the Brayton magnetic cycle;

– the AMR (active magnetic regenerative) cycle.

By comparison with the cycles of thermal machines, we can represent these magnetic refrigeration cycles with an (S,T) diagram (Figure 5.5).

5.2.3.1. *The magnetic Carnot cycle*

The magnetic Carnot cycle is an ideal cycle. The Carnot cycle is composed of two adiabatic processes and two isothermal processes:

– **Adiabatic magnetization** (1→2). The material is magnetized. This results in an increase in temperature.

– **Isothermal magnetization** (2→3). The temperature is kept constant. Heat is rejected at the hot reservoir.

– **Adiabatic demagnetization** (3→4). The magnetic field decreases. The temperature of the magnetocaloric material decreases.

– **Isothermal demagnetization** (4→1). The material is demagnetized. The temperature is constant. Heat is absorbed from the cold reservoir.

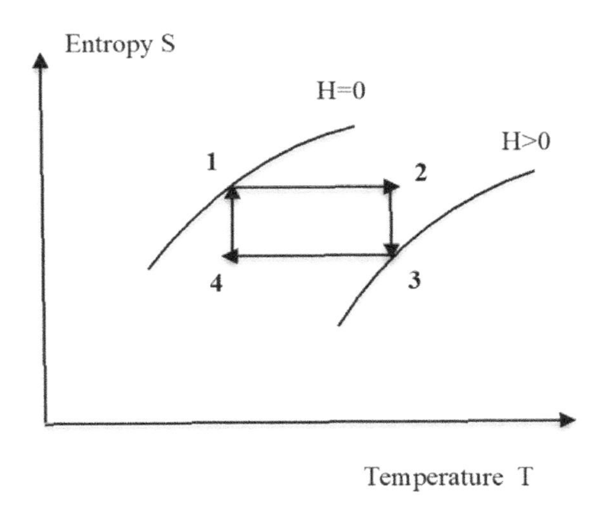

Figure 5.6. *Magnetic Carnot cycle*

5.2.3.2. *The magnetic Ericsson cycle*

The Ericsson cycle consists of two isothermal processes and two isofield processes:

– **Isofield heating** (1→2). The material heats up by isofield heating by taking heat from the heat transfer fluid.

– **Isothermal magnetization** (2→3). The material is magnetized and transfers heat to the hot reservoir while maintaining the same temperature.

– **Isofield cooling** (3→4). The material cools down by rejecting heat to the heat transfer fluid.

– **Isothermal demagnetization** (4→1). The material is demagnetized and absorbs heat from the cold reservoir.

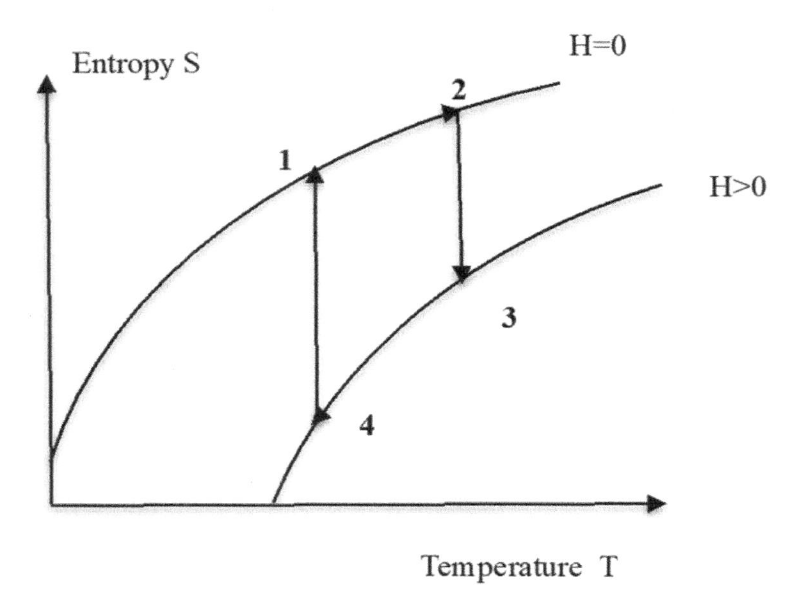

Figure 5.7. *Magnetic Ericsson cycle*

5.2.3.3. *The magnetic Brayton cycle*

The Brayton cycle is composed of isentropic magnetization and demagnetization with isofield cooling and heating:

– **Adiabatic magnetization** (1→2). The material is magnetized and its temperature increases.

– **Isofield cooling** (2→3). The material is cooled by isofield cooling by rejecting heat to the heat transfer fluid.

– **Adiabatic demagnetization** (3→4). The field applied to the material decreases and the temperature of the magnetocaloric material also decreases.

– **Isofield heating** (4→1). The material heats up by isofield heating by taking heat from the heat transfer fluid.

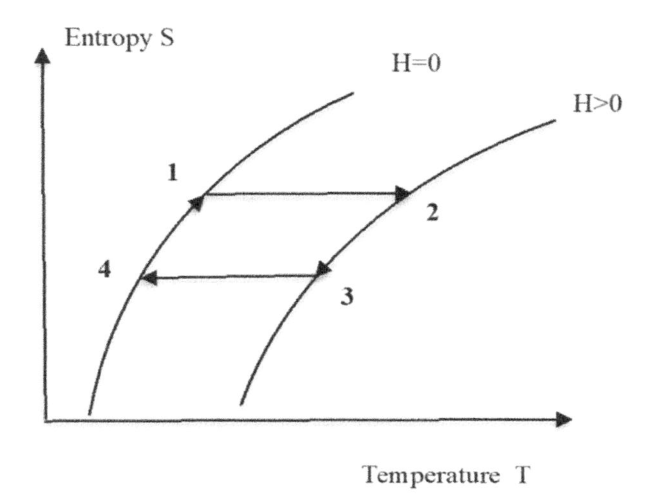

Figure 5.8. *Magnetic Brayton cycle*

5.2.3.4. *The AMR cycle*

To achieve large temperature differences, it is possible to use an AMR (active magnetic regeneration) cycle (Barclay and Steyert 1982). Indeed, the adiabatic temperature change (ΔT_{ad}) for a magnetic cycle is small. The value of ΔT_{ad} of gadolinium (the best known magnetocaloric material) in a magnetic field of 1 T (which can be obtained with permanent magnets) is about 3 K. Thus, to obtain a temperature difference comparable to conventional refrigeration, regeneration is an interesting solution (Barclay 1988).

AMR is a concept used in most magnetic refrigeration systems around room temperature (Lebouc et al. 2005).

An AMR cycle is composed of four main processes:

– **Adiabatic magnetization** (1→2). The material is magnetized and its temperature increases.

– **Isofield cooling** (2→3). Each slice of material is cooled by isofield cooling by rejecting heat to the heat transfer fluid.

– **Adiabatic demagnetization** (3→4). The field applied to the material decreases and the temperature of the magnetocaloric material decreases.

– **Isofield heating** (4→1). The material heats up by recovering heat from the heat transfer fluid.

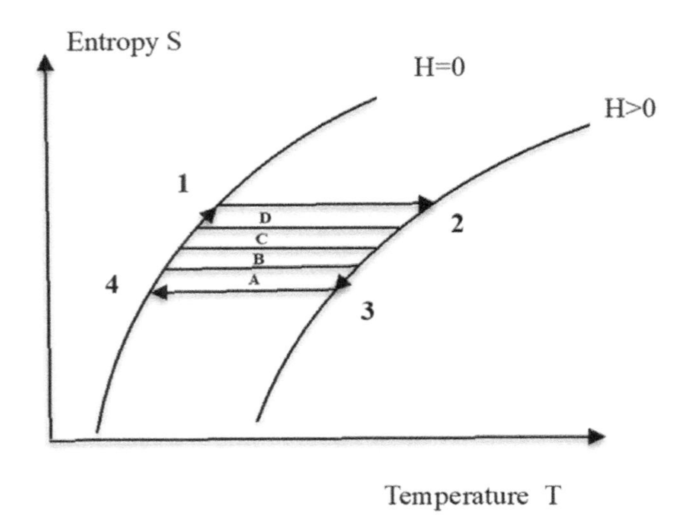

Figure 5.9. *Magnetic AMR cycle*

One way to represent the AMR cycle is to consider that each element of the magnetocaloric material undergoes a Brayton cycle, and that all of these elements are in series. The operation is similar to that of a multi-stage system.

Among the various existing magnetic cycles, the AMR cycle is considered the most promising for a magnetic refrigerator operating at room temperature (Aprea and Maiorino 2010; Greco et al. 2019).

5.2.4. *Magnetocaloric materials*

Magnetocaloric materials are a fundamental part of the thermodynamic performance of magnetic refrigerators. Room-temperature magnetic refrigeration applications require materials with a large MCE. By a large MCE, we mean high temperature variations (a few kelvins to a few tens of kelvins) for magnetic inductions accessible by magnets or superconducting electromagnets (a few teslas). The materials used are rare earth metals, either pure or combined as alloys.

Magnetocaloric materials are classified into two categories:

– first-order materials, which have a magnetization that changes abruptly with temperature;

– second-order materials, for which this variation is softer.

First-order materials are characterized by large entropy variations, but over a small temperature range. Second-order materials have small entropy variations, but over a wide temperature range.

A magnetocaloric material must have several characteristics (Lionte 2015):

– a Curie temperature close to the operating temperature;

– a significant variation in temperature and entropy for basic values of the magnetic field intensity;

– a wide range of entropy and temperature variations;

– high thermal conductivity to improve the exchange efficiency;

– minimal magnetic and thermal hysteresis to allow high operating frequencies and high power;

– high electrical resistance to reduce Eddy current losses;

– interesting mechanical properties for the manufacture and operation;

– good technical and economic characteristics (cost, abundance and availability).

The most widely used material for magnetic refrigerator prototypes is gadolinium, the only magnetic material available in a pure state with a giant MCE. The values of ΔT_{ad} are approximately 2.8 K/T for a magnetic field of 1 T.

Figure 5.10. *Gadolinium regenerator (Legait et al. 2014)*

Because of the high price and low availability of gadolinium, research is now focusing on alloys based on lanthanum–iron–silicon (LaFeSi) or manganese such as MnFePAs or MnFePSi.

These alloys have better magnetocaloric performances than gadolinium. Figure 5.11 shows the performance of different magnetocaloric materials as a function of a) the number of transfer units (NTU), b) thermal conductivity and c) frequency (Legait et al. 2014).

Figure 5.11. *Heat exchanger made up of magnetocaloric composite materials based on La(Fe1-xSix)13 (Skokov et al. 2014)*

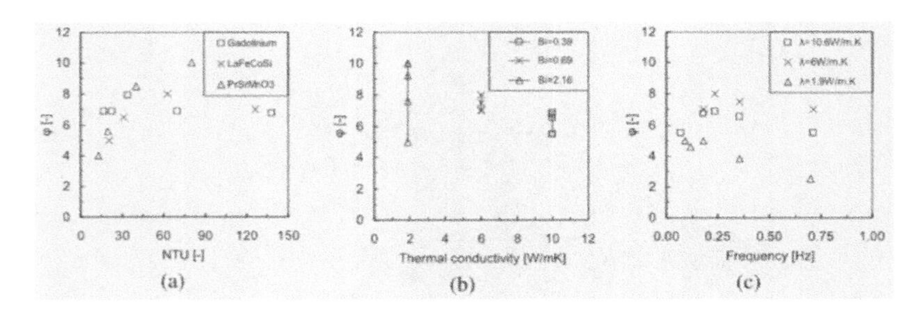

Figure 5.12. *Performance of different magnetocaloric materials as a function of a) the number of transfer units (NTU), b) thermal conductivity and c) frequency (Legait et al. 2014)*

A recent overview concerning magnetocaloric materials has been proposed by Mayer (2017). Figure 5.12 shows the properties of magnetocaloric materials for a compromise between maximizing the performance (high magnetic entropy variation) and minimizing the cost (Mayer et al. 2017).

The environmental impact of magnetocaloric materials studied by Mayer (2017) is presented in Figure 5.13. Mayer defines the environmental impact of a material as follows: "The environmental impact of a material can be characterized by the quantity of CO_2 emitted and water consumed to obtain 1 kg of material by transformation of the raw material." Note in Figure 5.13 that magnetocaloric materials have a relatively low environmental impact (Mayer et al. 2017).

Mayer identifies three La (families of magnetocaloric materials that are most promising today) (Mayer et al. 2017):

– the manganite family;

– the La(Fe,Si)13 family;

– the family of pnictides Mn2 – xFex(P1 – ySiy).

These three families of magnetocaloric materials have interesting properties. However, it should be noted that the MCE introduces structural modifications and a thermal hysteresis, which is a source of irreversibility.

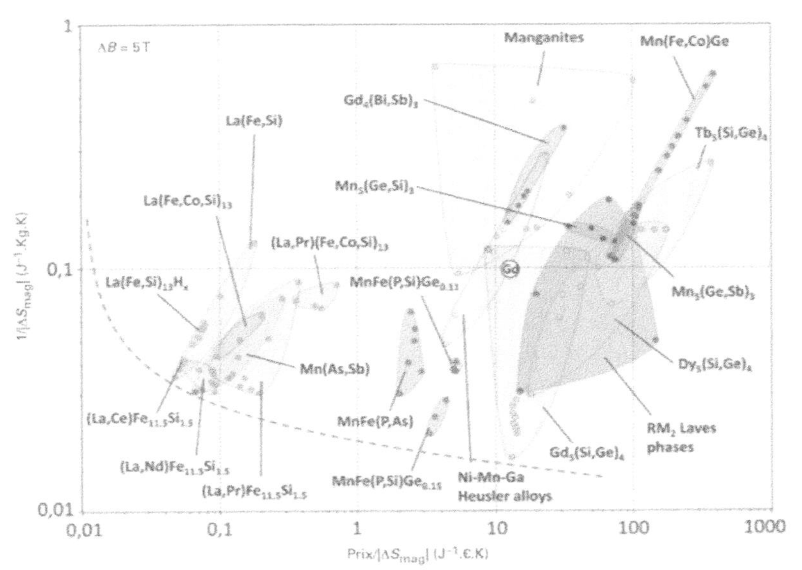

This diagram was created using the software CES Constructor 2016 (Granta Design). The database of magnetocaloric matrials is available on request (contact S. Gorsse).

Figure 5.13. *Properties of magnetocaloric materials (Mayer et al. 2017)*

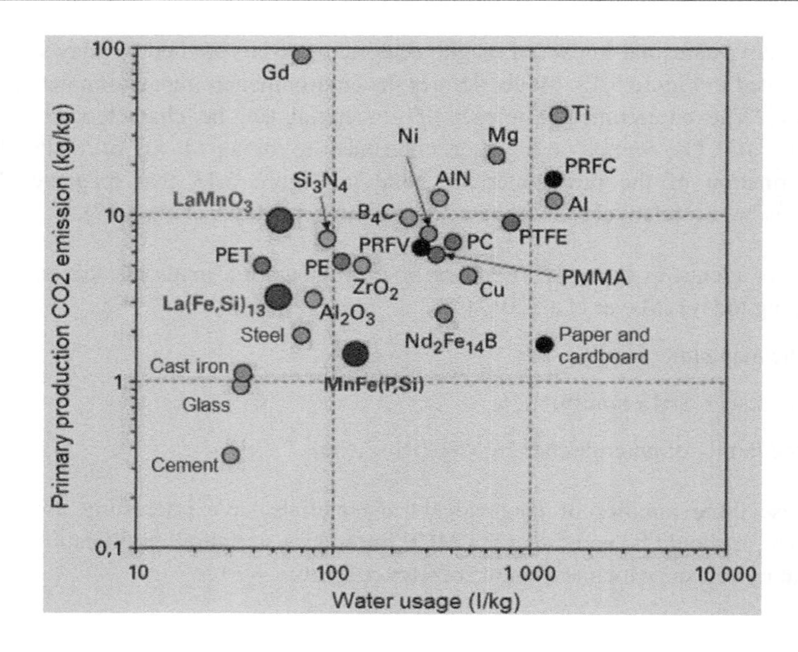

Figure 5.14. *Amount of CO_2 emitted and water used per kg of material produced per transformation of global annual production (Mayer et al. 2017)*

5.3. Numerical models

5.3.1. *Numerical models of magnetocaloric regenerators*

The regenerator is the main element of a magnetic refrigerator (Yao et al. 2006). The regenerator is made of a magnetocaloric material, often gadolinium (Balli et al. 2012) or alloys of magnetocaloric materials.

The typical geometries of the regenerators are the following: plates, powders, balls, metallic grids. They are shown in Figure 5.15. The design criteria for regenerators are as follows: the ratio between the useful surface and the volume of the material, the material properties and the manufacturing constraints.

The performance of AMRs is highly dependent on the thermal performance of the regenerative matrix. Trevizoli et al. (2017) compared the performance of three different regenerator geometries: parallel plate, lattice and packed bed of spheres. The best performances were obtained for the spheres and the lattice (Figure 5.16).

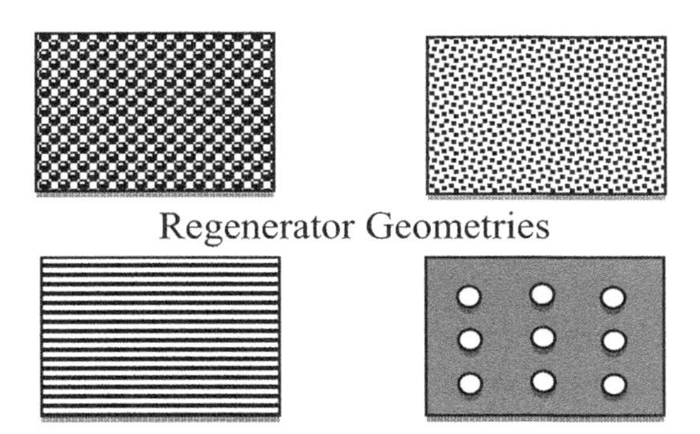

Figure 5.15. *Different geometries of magnetic regenerators*

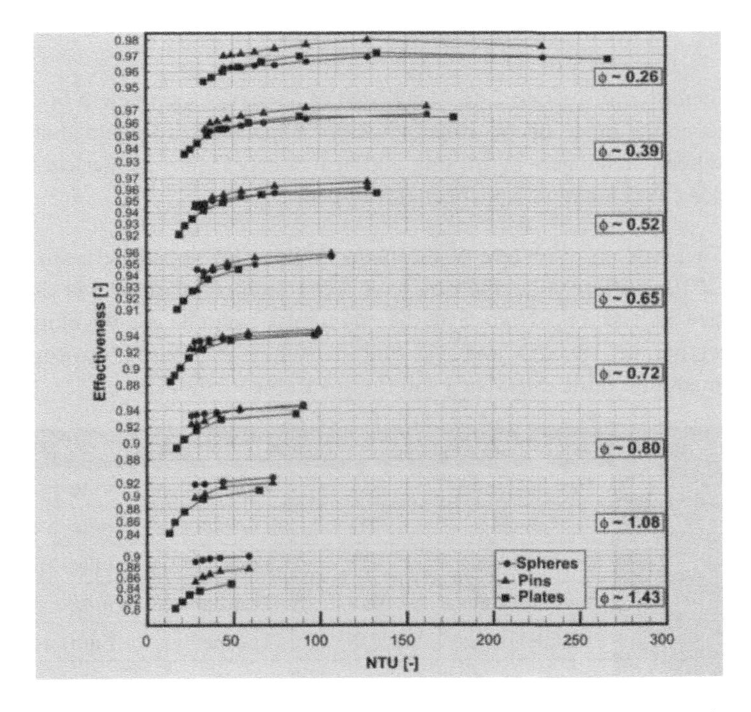

Figure 5.16. *Performance for different configurations of magnetic regenerators according to the NTU (Trevizoli et al. 2017)*

In order to improve their efficiency, these regenerators can be associated in series or in parallel (Figure 5.17), depending on whether we wish to increase the temperature difference or the cooling capacity (Roudaut 2011).

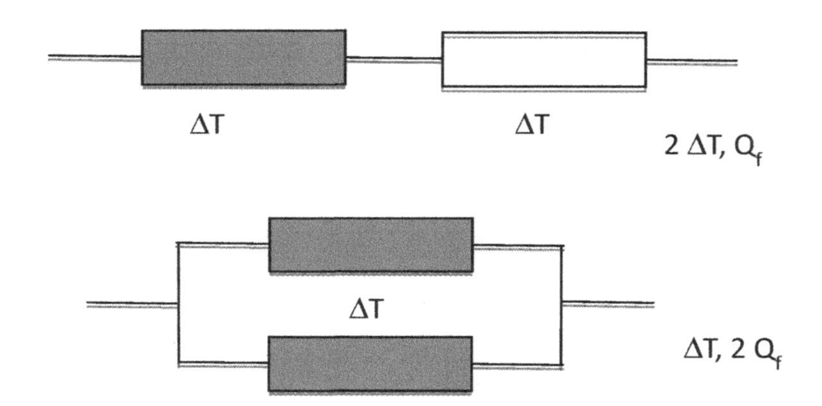

Figure 5.17. *Series and parallel magnetic regenerators*

Magnetocaloric refrigerators are composed of a magnet (permanent or electric), a porous regenerator made up of a magnetocaloric material, a system that pumps a coolant and heat exchangers at the sink and heat source. Many numerical models of magnetocaloric regenerators have been developed.

Table 5.1 gives an overview of the main models developed between the years 2000 and 2010. Different geometries have been studied: parallel plates, spheres, porous lattice and layers. Note that the numerical models are developed in the transient regime in 1D, 2D and 3D and that they are based on conventional numerical methods (finite difference and finite volume methods).

Author	Geometry	Dimension	Regime	Numerical method
Allab et al. (2005)	Parallel plates	1D	Transient	Finite difference
Kitanovski (2005)	Layers	1D	Transient	Finite difference
Li et al. (2006)	Porous lattice	1D	Transient	Finite difference
Petersen et al. (2008)	Parallel plates	2D	Transient	Finite difference
Nielsen et al. (2009)	Parallel plates	2D	Transient	Finite volume
Bouchard et al. (2009)	Spheres	3D	Transient	Finite volume

Table 5.1. *Numerical models of magnetocaloric regenerators*

These models have made it possible to optimize the geometric configurations and operating modes of magnetocaloric regenerators (Bahl et al. 2014). The majority of models use the AMR cycle, which is most effective for optimizing heat transfer.

5.3.2. *Recent numerical models*

Several models have been developed recently, including Almanza (2014), Lionte (2015), You et al. (2017) and Plait (2019). The model developed by Lionte (2015) during his thesis at the INSA Strasbourg ICUBE laboratory, which has been the subject of several publications (Lionte et al. 2014a, 2014b, 2014c), is discussed here.

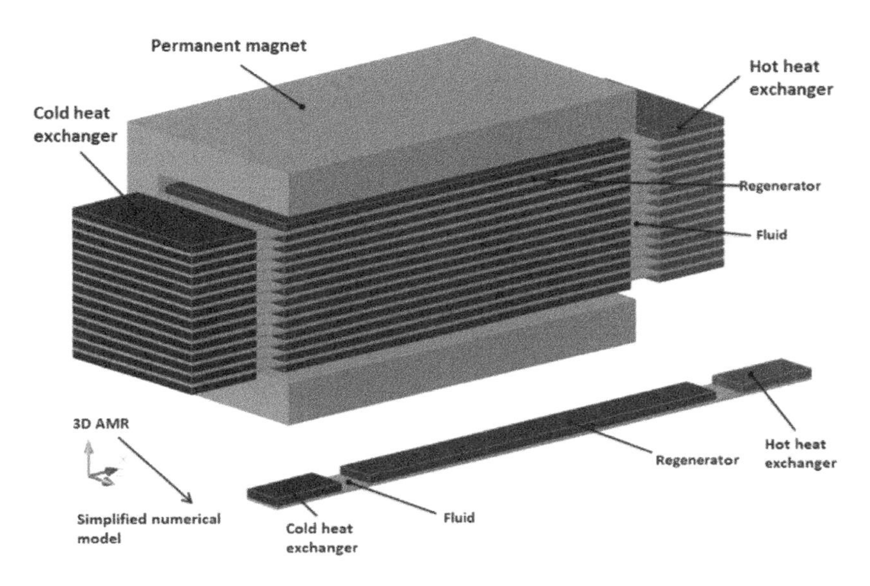

Figure 5.18. *Geometry of the Lionte model (Lionte 2015). For a color version of this figure, see www.iste.co.uk/bonjour/refrigerators.zip*

Lionte modeled a parallel-plate alternating AMR that operates at room temperature. The model is a 2D unsteady-state numerical model developed on COMSOL. This model makes it possible to evaluate the performance of the system: evolution of the temperature gradient in the regenerator during a cycle, evolution of the temperature for an AMR, cold power and COP. The model was validated with experimental results provided by a project partner (Lionte 2015).

The model geometry is shown in Table 5.2. It is a regenerator made up of 14 flat parallel gadolinium plates. The heat transfer fluid used is water. On either side of the regenerator are the cold (CHEX) and hot (HHEX) heat exchangers. The operating characteristics are shown in Table 5.3. The simulated magnetic field is between 1 T and 2 T. The AMR uses an alternating movement of the fluid with a speed between

0.025 m/s and 0.15 m/s and a frequency of motion of the magnets between 0.3 and 2 Hz in a cyclic process (Lionte et al. 2014a, 2014b, 2014c).

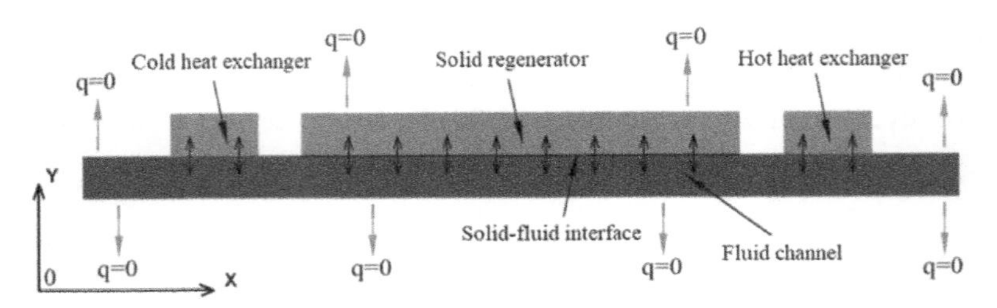

Figure 5.19. *Boundary conditions of the numerical model (Lionte 2015)*

Quantity	Values (mm)
Channel length	60–200
Regenerator length	30–100
Length of exchangers	5
Channel height	0.15–0.60
MMC plate thickness	0.3–1.00
Material width	10
Regenerator porosity	0.218

Table 5.2. *Geometric quantities of the Lionte model (Lionte 2015)*

Quantity	Values
Magnetic field	1.0–2.0 T
Frequency	0.3–2.0 s^{-1}
Fluid velocity	0.025–0.15 m/s
Initial system temperature	294.45 K
Time interval	0.1 s

Table 5.3. *Features of the Lionte operation (Lionte 2015)*

Figure 5.19 presents the boundary conditions of the numerical model. The MCE is taken into account by introducing a variable source term in the conservation of energy equation. The energy conservation equations in the solid and the fluid, as well as the momentum conservation equation in the fluid are solved in Comsol using the finite element method. The model was validated by comparing the numerical results with the experimental results (Figure 5.21).

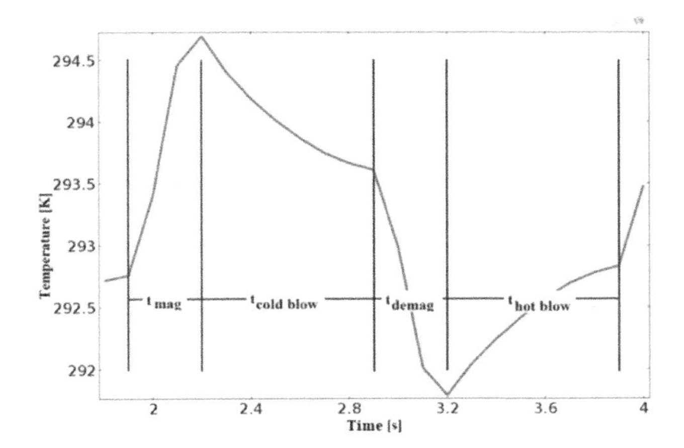

Figure 5.20. *Regenerator temperature during a cycle (Lionte et al. 2014a, 2014b, 2014c)*

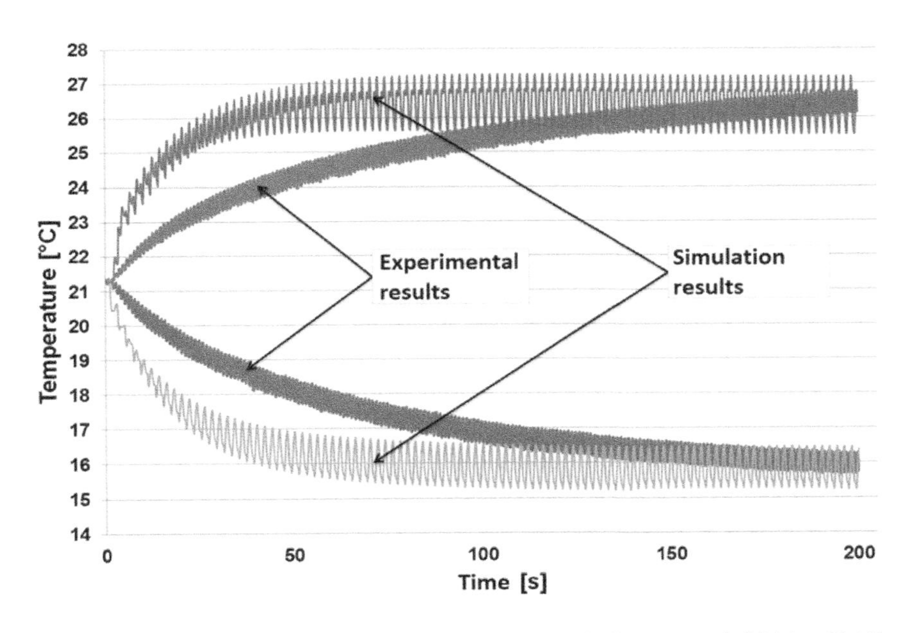

Figure 5.21. *Evolution of the temperature for an AMR (Lionte et al. 2014a, 2014b, 2014c). For a color version of this figure, see www.iste.co.uk/bonjour/refrigerators.zip*

Figure 5.20 shows the temperature in the regenerator during a cycle. Figure 5.21 presents the evolution of the temperature for the AMR. After an operating cycle, the regenerator enters a steady state. We note that in a steady state, we have a good model–measurement correlation.

The 2D distribution of temperatures during a cycle is shown in Figure 5.22.

Figure 5.23 shows the cooling capacity as a function of the temperature difference between the heat exchangers. The cooling capacity reaches maximum values at a temperature difference of 0.5–4°C.

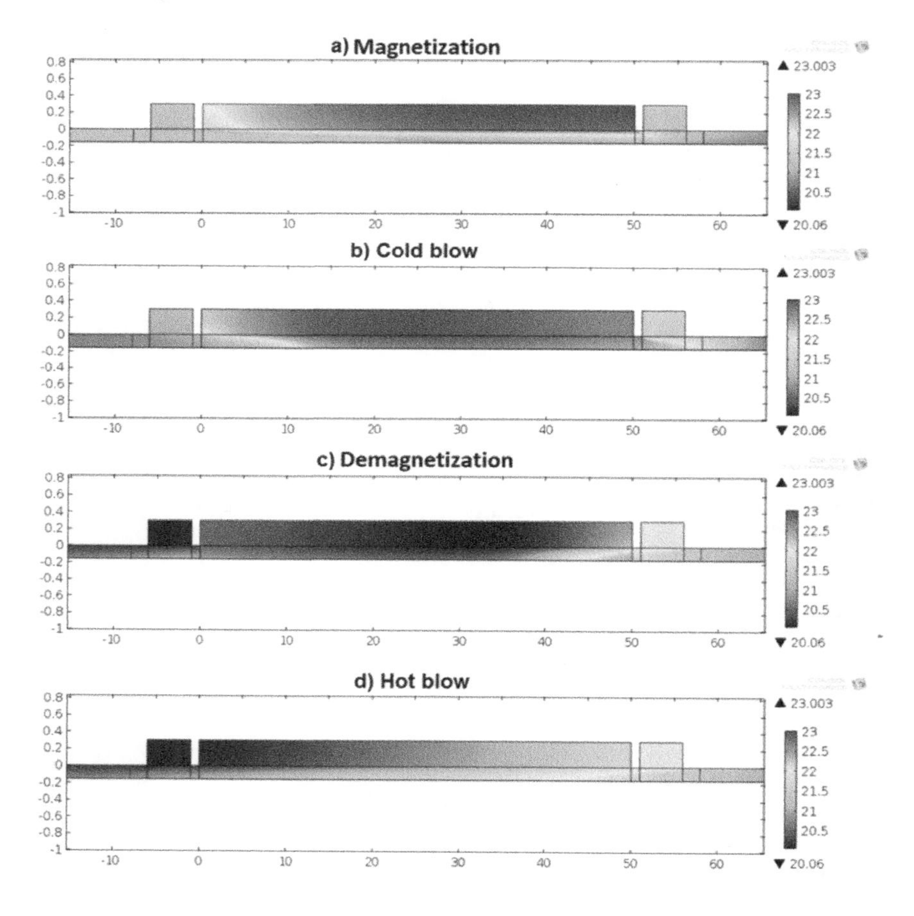

Figure 5.22. *The 2D distribution of temperatures during a cycle (Lionte 2015). For a color version of this figure, see www.iste.co.uk/bonjour/refrigerators.zip*

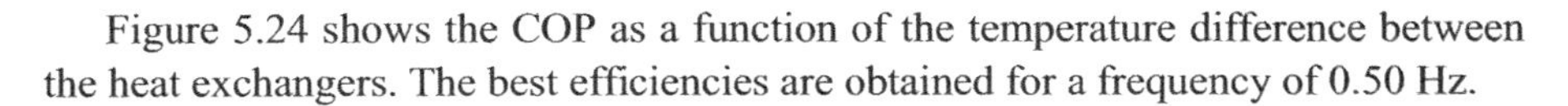

Figure 5.24 shows the COP as a function of the temperature difference between the heat exchangers. The best efficiencies are obtained for a frequency of 0.50 Hz.

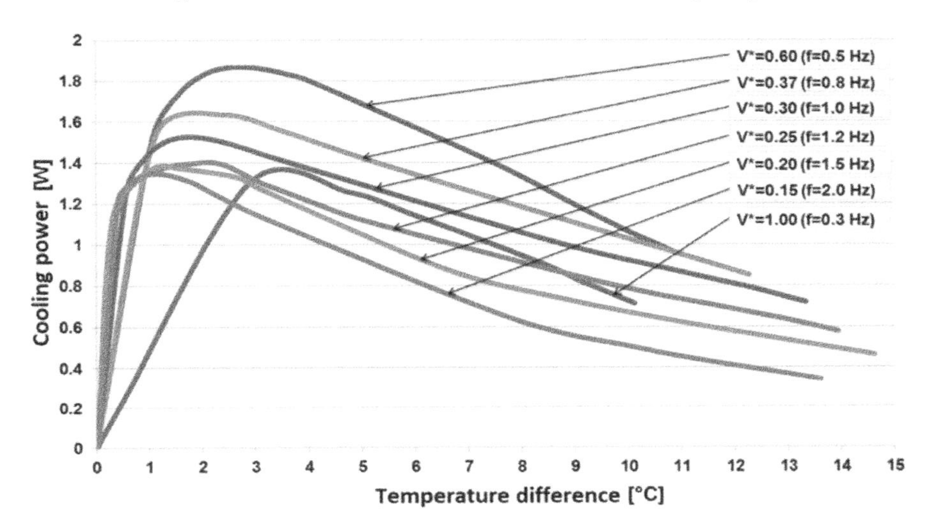

Figure 5.23. *Cooling power as a function of temperature difference (Lionte 2015). For a color version of this figure, see www.iste.co.uk/bonjour/refrigerators.zip*

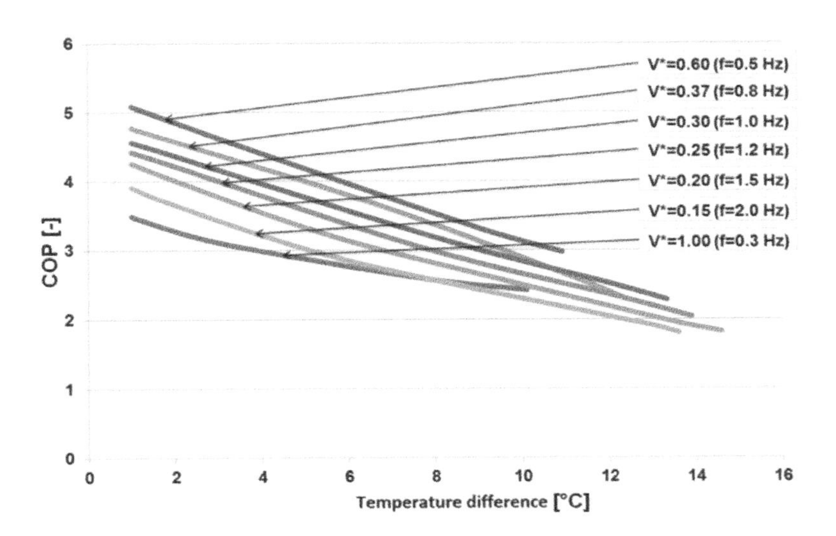

Figure 5.24. *Coefficient of performance (Lionte 2015). For a color version of this figure, see www.iste.co.uk/bonjour/refrigerators.zip*

5.4. Applications

5.4.1. *Prototypes*

There are numerous prototypes of magnetic refrigeration around ambient temperature. In 1976, Brown developed the first passive regenerator for room-temperature magnetic refrigeration applications (Brown 1976). The prototype worked with superconducting magnets (B = 7 T) and the regenerator was composed of gadolinium parallel plates.

Zimm et al. (1998) developed a magnetic refrigerator with superconducting magnets in 1998 and the first rotating magnetocaloric refrigeration with permanent magnets in 2001 (Zimm et al. 2001). This was an important advance in the field of magnetic refrigeration. It was the first prototype with interesting characteristics: 50 W cooling capacity and 25 K temperature difference.

In 2014, Victoria University presented a new rotary prototype (Arnold et al. 2014). The device has two AMRs filled with spherical gadolinium particles, with a maximum cooling capacity of 2.5 W and a temperature difference of 33 K.

Jacobs developed a prototype for a large-scale rotating magnetic refrigerator, characterized by a cooling capacity of 2,000 W and an electrical coefficient of performance (COPe) greater than 2 (Jacobs et al. 2014).

Johra's (2019) prototype is an active regeneration rotating magnetic system. The MCM, in the form of spheres, is compacted inside trapezoidal cassette regenerators. With gadolinium, the performances of this prototype are a heating power of 2,600 W, a COP of 3.93 and a maximum temperature difference of approximately 20 K.

Nakamshima's et al. (2021) prototype enables temperature control of a refrigerator for wine bottles. The main components of the system are an AMR composed of eight porous beds of Gd/Gd–Y spheres, a magnetic circuit, a set of solenoid valves for flow control and tube fin heat exchangers. The system can reach an average temperature of 10.8°C inside the wine cabinet for an ambient temperature of 25°C. The COP is approximately 0.38.

Magnetocaloric prototypes can be classified based on:

– the type of magnet used (electromagnet or permanent magnet);

– the morphology of the regenerator (spheres or plates);

– the type of heat transfer fluid (gas or liquid);

– the technology.

From a technological point of view, prototypes can be classified into linear prototypes and rotary prototypes (Kitanovski 2020).

Linear prototypes are characterized by intense fields and low power (a few tens of watts). Linear-type devices are mainly used for testing AMRs and consist of permanent magnet assemblies (Figure 5.25).

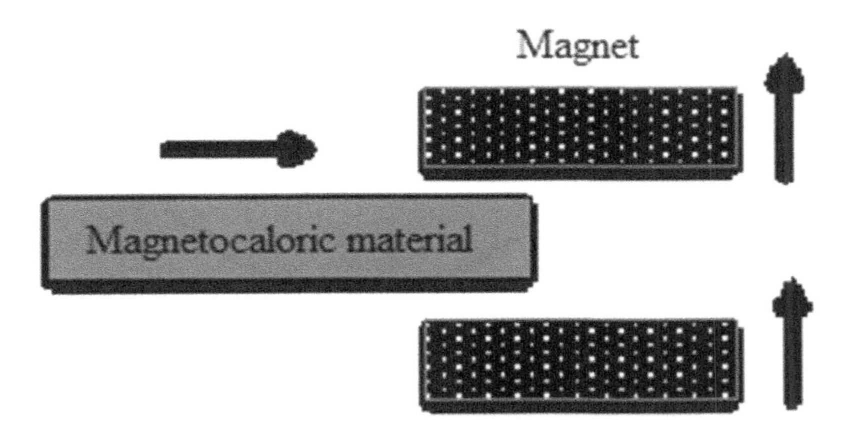

Figure 5.25. *Linear prototype*

Rotary prototypes are more compact and characterized by higher powers (several hundred watts) and can operate up to frequencies of 10–15 Hz. Most prototypes are rotary (Figure 5.25).

In recent years, approximately 30 linear prototypes and approximately 30 rotary prototypes have been developed mainly in the USA, Denmark, Canada, China, Japan, Korea, France, Slovenia, Switzerland, Germany, the United Kingdom, Brazil and Spain (Yu et al. 2010; Greco et al. 2019). Table 5.4 presents some examples of prototypes as well as their characteristics and efficiencies. We note that the most widely used magnetocaloric material is gadolinium. The temperature difference obtained is greater than 25 K and the maximum cooling capacity is approximately 2,000 W.

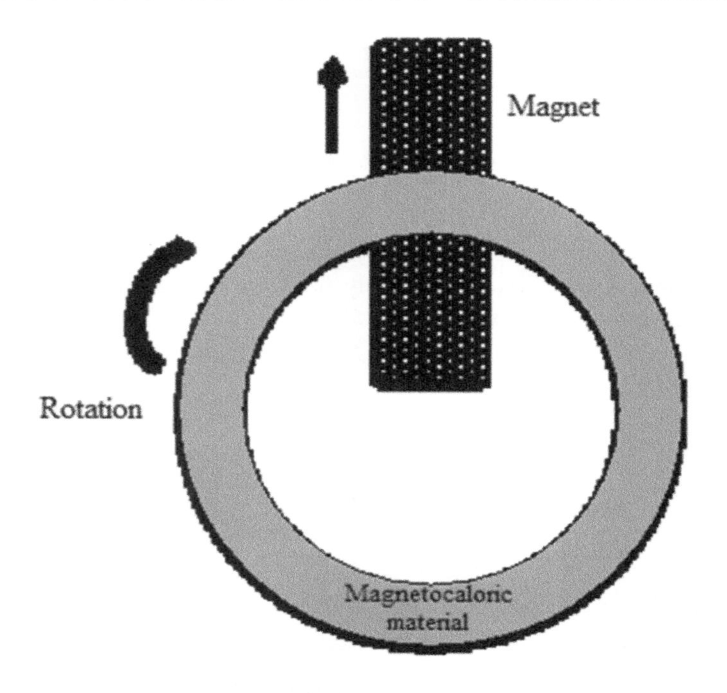

Figure 5.26. *Rotary prototype*

Prototype	Type	Magnetic field	ΔT (K)	Regenerator	Capacity (W)
Brown (1976)	Linear	Superconductor B = 7 T	47	Gadolinium plates	–
Zimm et al. (2001)	Rotary	Permanent magnets B = 1.5 T	25	Gadolinium spheres	50
Arnold (2014)	Rotary	Permanent magnets B = 1.5 T	33	Gadolinium spheres	2.5
Jacobs (2014)	Rotary	NdFe magnets B = 1,44 T	11	LaFeSiH particles	2,000
Johra (2018)	Rotary	Permanent magnets B = 1.46 T	20	Gadolinium	2,600
Nakashima et al. (2021)	Rotary	Permanent magnets B = 1 T	10	Gd/Gd–Y spheres	28

Table 5.4. *Prototypes at ambient temperature*

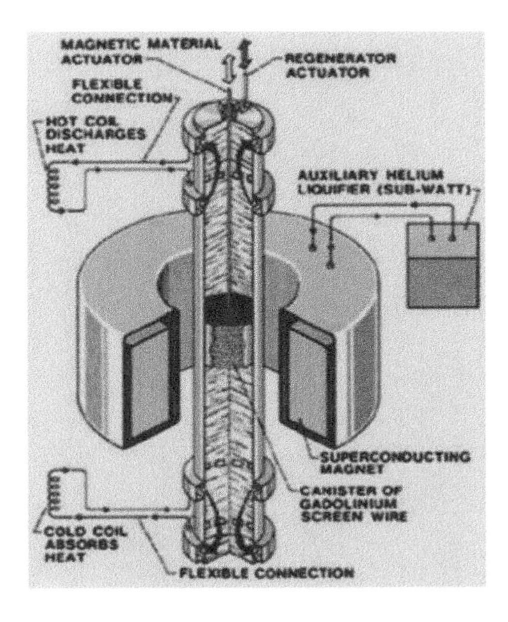

Figure 5.27. *Brown's (1976) prototype*

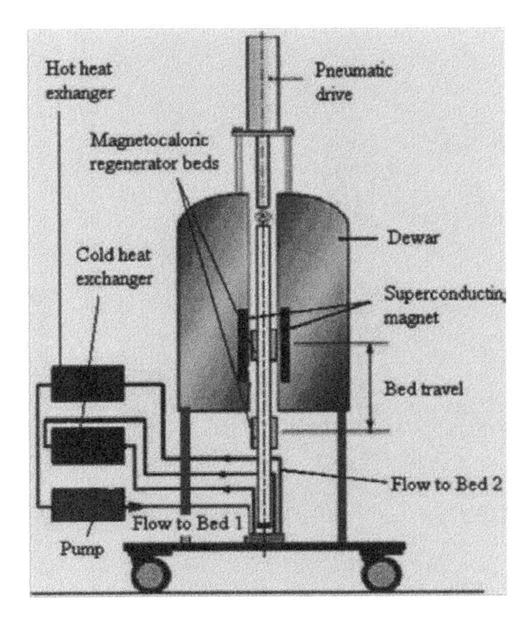

Figure 5.28. *Zimm's (1998) prototype*

Figure 5.29. *Arnold's (2014) prototype. For a color version of this figure, see www.iste.co.uk/bonjour/refrigerators.zip*

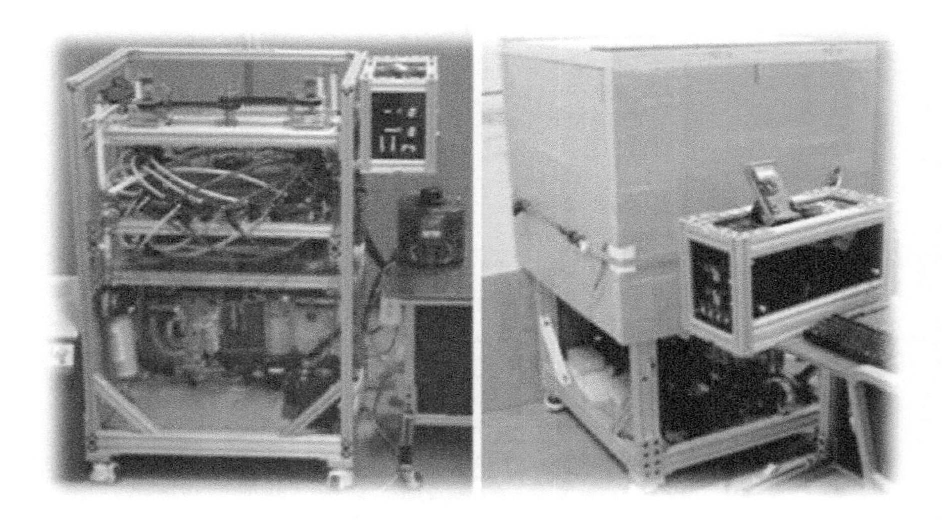

Figure 5.30. *Jacobs' (2014) prototype*

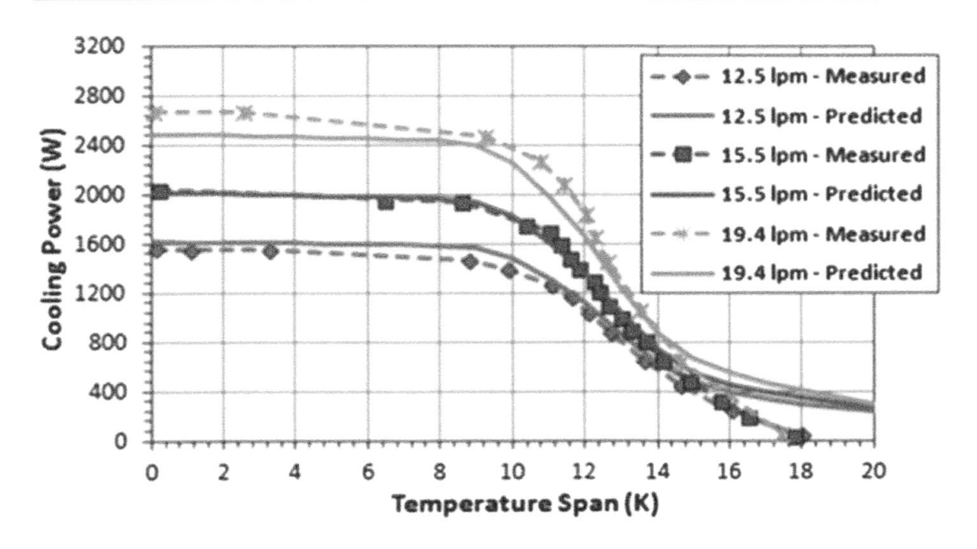

Figure 5.31. *Cooling power – Jacobs' (2014) prototype*

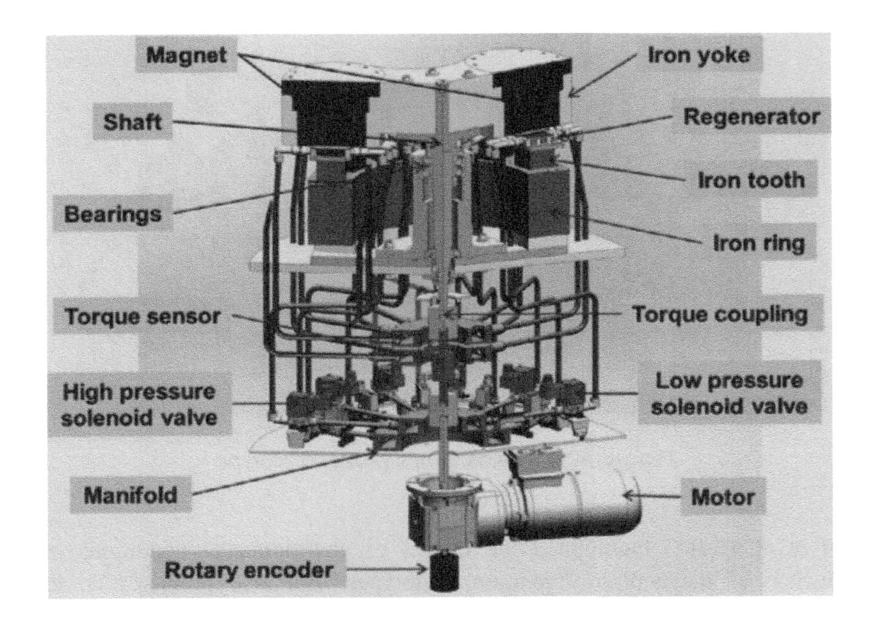

Figure 5.32. *Johra's (2019) prototype*

Figure 5.33. *Nakashima's (2021) prototype*

Wu et al. (2014) highlighted the impact of environmental conditions on the performance of magnetic refrigeration prototypes. Wu also proposed an exergy analysis of magnetic refrigeration prototypes to assess the performance of the prototypes.

Figure 5.34 shows the exergy cooling power as a function of the temperature difference for:

– the prototype developed by the University of Victoria in Canada (Tura and Rowe 2011);

– the prototype developed by the Technical University of Denmark (Lozano et al. 2014).

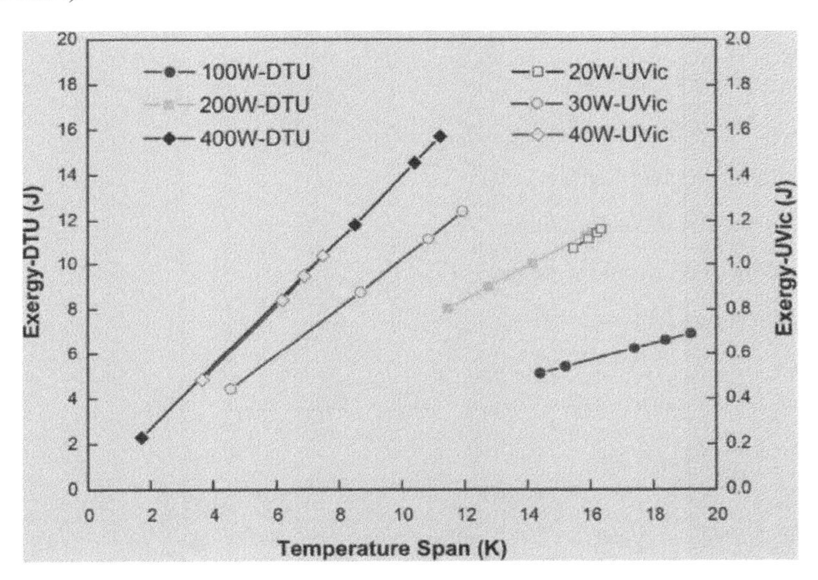

Figure 5.34. *Exergy-equivalent cooling power as a function of the temperature difference for the prototypes developed by the University of Victoria (UVic) in Canada (Tura and Rowe 2011) and by the Technical University of Denmark (DTU) (Lozano et al. 2014)*

5.4.2. *Future applications*

The advantages of magnetic refrigeration are:

– a COP superior to conventional refrigeration techniques;

– the absence of atmospheric pollutants;

– the absence of noise;

– compactness.

This technology has drawbacks that cannot be ignored, which requires:

– high-performance (high magnetic entropy variation) and expensive magnetocaloric materials;

– strong magnetic fields;

– high manufacturing costs.

The potential applications of magnetic refrigeration are numerous:

– domestic refrigeration;

– industrial refrigeration;

– building air conditioning;

– automotive air conditioning;

– cooling of electronic systems;

– applications for spaceships.

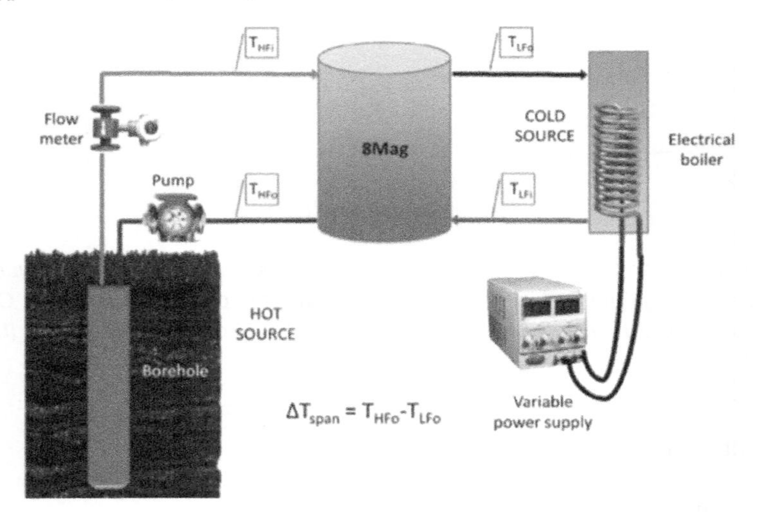

Figure 5.35. *Principle of a GeoThermag geothermal magnetic refrigerator (Aprea et al. 2015)*

Figure 5.35 shows the GeoThermag geothermal magnetic refrigerator concept introduced by Aprea (2015). It is a magnetic refrigerator connected to a geothermal probe. The advertised efficiencies are a cooling capacity of 190 W and a maximum COP of 2.20.

In 2018, Johra (2018) presented a more efficient geothermal magnetic refrigerator (Figure 5.36), characterized by a maximum heating power of 2,600 W and a maximum COP of approximately 10 (Figure 5.37).

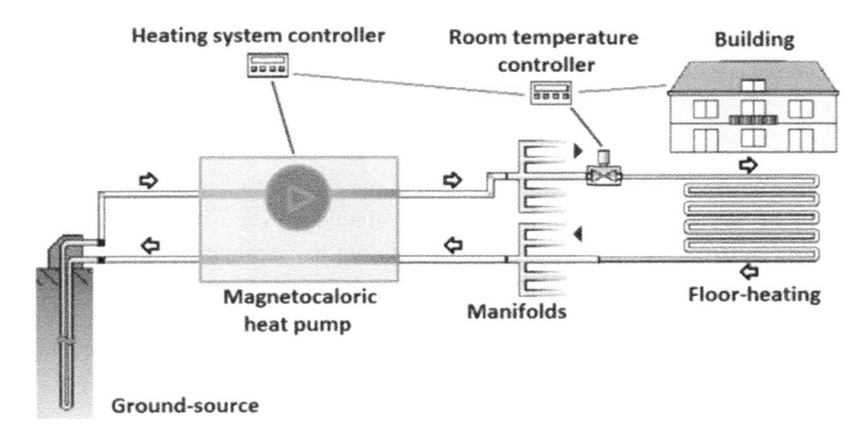

Figure 5.36. *Integration of a magnetocaloric heat pump in a geothermal system composed of a geothermal heat exchanger and floor heating (Johra et al. 2019)*

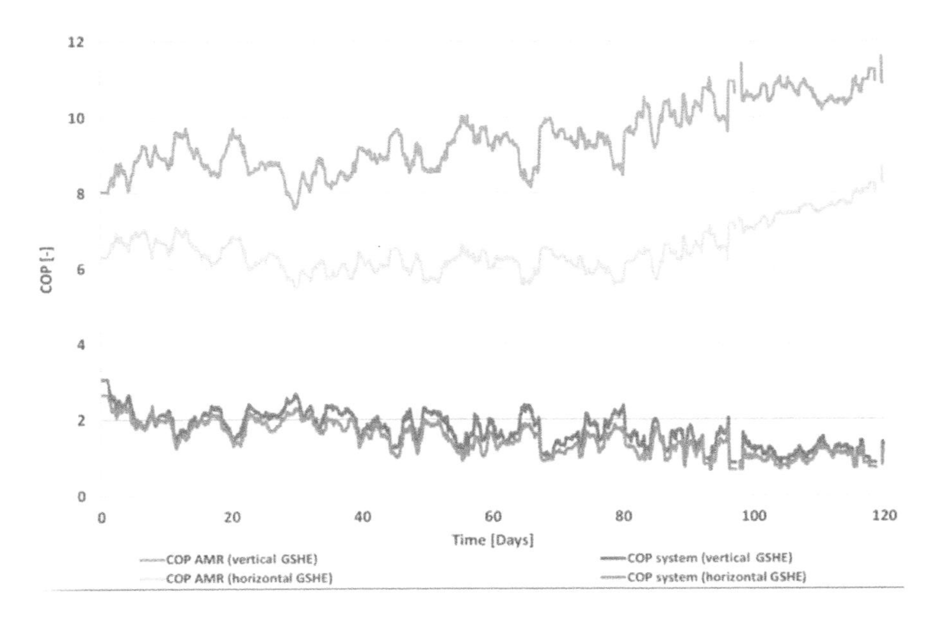

Figure 5.37. *COP of the magnetocaloric heat pump in a geothermal system (Johra 2018)*

Some studies are devoted to the development of magnetic refrigeration for the automotive industry (Kieffer 2012). Such systems must operate at high frequencies due to reduced size and weight.

Recently, a prototype for a magnetic wine cooler was developed (Nakashima 2021). The maximum COP obtained was 0.38.

5.5. Conclusion

Magnetic refrigeration, a still emerging technology, can, in some cases, be a sustainable alternative to refrigeration due to its economic and ecological advantages. Despite this, magnetic refrigeration is not yet competitive with conventional refrigeration technologies. To achieve industrialization, several challenges must be met: designing better magnetocaloric materials, developing strong magnetic fields, optimizing systems and manufacturing costs.

5.6. References

Allab, F., Kedous-Lebouc, A., Fournier, J., Yonnet, J. (2005). Numerical modeling for active magnetic regenerative refrigeration. *IEEE Trans. Magn.*, 41(10), 3757–3759.

Almanza, M. (2014). Réfrigération magnétique : conceptualisation, caractérisation et simulation. PhD Thesis, Université de Grenoble.

Almanza, M., Pasko, A., Mazaleyrat, F., Lo Bue, M. (2017). Numerical study of thermomagnetic cycle. *J. Magn. Magn. Mater.*, 426, 64–69.

Aprea, C. and Maiorino, A. (2010). A flexible numerical model to study an active magnetic refrigerator for near room temperature applications. *Appl. Energy*, 87(8), 2690–2698

Aprea, C., Greco, A., Maiorino, A. (2015). GeoThermag: A geothermal magnetic refrigerator. *Int. J. Refrigeration*, 59, 75–83.

Arnold, D.S., Tura, A., Rubesatt-Trott, A., Rowe, A. (2014). Design improvements of a permanent magnet active magnetic refrigerator. *Int. J. Refrigeration*, 37, 99–105.

Bahl, C.R.H., Engelbrecht, K., Eriksen, D., Lozano, J.A., Bjørk, R., Geyti, J., Nielsen, K.K., Smith, A., Pryds, N. (2014). Development and experimental results from a 1 kW prototype AMR. *Int. J. Refrigeration*, 37, 78–83.

Balli, M., Sari, O., Mahmed, C., Besson, C., Bonhote, P., Duc, D., Forchelet, J. (2012). A pre-industrial magnetic cooling system for room temperature application. *Appl. Energy*, 98, 556–561.

Barclay, J.A. (1988). Magnetic refrigeration – A review of a developing technology. *Adv. Cryog. Eng.*, 33, 719–731.

Barclay, J.A. and Steyert, W.A. (1982). Active magnetic regenerator. 4, 332,135 U.S. Patent.

Bouchard, J., Nesreddine, H., Galanis, N. (2009). Model of a porous regenerator used for magnetic refrigeration at room temperature. *Int. J. Heat Mass Transf.*, 52(5–6), 1223–1229.

Bouchekara, H. (2008). Recherche sur les systèmes de réfrigération magnétique : modélisation numérique, conception et optimisation. PhD Thesis, Institut National Polytechnique de Grenoble.

Brown, G.V. (1976). Magnetic heat pumping near room temperature. *J. Appl. Phys.*, 47, 3673–3680.

Brown, G.V. (1978). Practical and efficient magnetic heat pump. *NASA Tech. Brief*, 3, 190–191.

Edison, T. (1892). Pyromagnetic generator. U.S. Patent No. 476,983.

Greco, A., Aprea, C., Maiorino, A., Masselli, C. (2019). A review of the state of the art of solid-state caloric cooling processes at room-temperature before. *Int. J. Refrigeration*, 106, 66–88.

Jacobs, S., Auringer, J., Boeder, A., Chell, J., Komorowski, L., Leonard, J., Russek, S., Zimm, C. (2014). The performance of a large-scale rotary magnetic refrigerator. *Int. J. Refrigeration*, 37, 84–91.

Johra, H. (2018). Integration of a magnetocaloric heat pump in energy flexible buildings. PhD Thesis, Aalborg University.

Johra, H., Filonenko, K., Heiselberg, P., Veje, C.T., Dall'oio, S., Engelbrecht, K., Bahl, C.R.H. (2019). Integration of a magnetocaloric heat pump in an energy flexible residential building. *Renew. Energy*, 136(115), 126

Kieffer, C. (2012). Conception optimale d'un système de refroidissement magnétocalorique à actionneur intégré : application à la climatisation automobile. PhD Thesis, Université de Franche-Comté.

Kitanovski, A. (2020). Energy applications of magnetocaloric materials. *Adv. Energy Mater.* 10, 1903741.

Kitanovski, A., Egolf, P.W., Gendre, F., Sari, O., Besson, C. (2005). A rotary heat exchanger magnetic refrigerator. In *First International Conference on Magnetic Refrigeration at Room Temperature*, Egolf, P.W. (ed.). International Institute of Refrigeration, Paris.

Lebouc, A., Allab, F., Fournier, J.M., Yonnet, J.P. (2005). Refrigération magnétique. *Technique de l'ingénieur*, 1–16

Legait, U., Guillou, F., Kedous-Lebouc, A., Hardy, V., Almanza, M. (2014). An experimental comparison of four magnetocaloric regenerators using three different materials. *Int. J. Refrigeration*, 37, 147–155

Li, P., Gong, M., Yao, G., Wu, J. (2006). A practical model for analysis of active magnetic regenerative refrigerators for room temperature applications. *Int. J. Refrigeration*, 29, 1259–1266.

Lionte, S. (2015). Caractérisation, étude et modélisation du comportement thermomagnétique d'un dispositif de réfrigération magnétique à matériaux non linéaires et point de Curie proche de la température ambiante. PhD Thesis, Université de Strasbourg.

Lionte, S., Vasile, C., Siroux, M. (2014a). Numerical analysis of a reciprocating active magnetic regenerator. *Appl. Thermal Eng.*, 75, 871–879.

Lionte, S., Vasile, C., Siroux, M. (2014b). Approche multi-physique et multi-échelle d'un régénérateur magnéto-thermique actif. *Proceeding Congrès SFT*, Lyon.

Lionte, S., Vasile, C., Siroux, M. (2014c). La réfrigération magnétique : technologie innovante de refroidissement pour des applications autour de la température ambiante. *Proceedings, Colloque Francophone en énergie COFRET*, Paris.

Lozano, J., Engelbrecht, K., Bahl, C., Nielsen, K.K., Barbosa Jr., J.R., Prata, A.T., Pryds, N. (2014). Experimental and numerical results of a high frequency rotating active magnetic refrigerators. *Int. J. Refrigeration*, 37, 92–98.

Mayer, C., Miraglia, S., Gorsse, S. (2017). Matériaux magnétocaloriques. *Techniques de l'ingénieur*, 1–20

Nakamshima, A.T.D., Fortkamp, F., De Sa, M., Dos Santos, V., Hoffmann, G., Peixer, G., Dutra, C., Ribeiro, M., Lozano, J., Barbosa Jr., J. (2021). A magnetic wine cooler prototype. *Int. J. Refrigeration*, 122, 110–121.

Nielsen, K.K., Bahl, C.R.H., Smith, A., Bjørk, R., Pryds, N., Hattel, J. (2009a). Detailed numerical modeling of a linear parallel-plate active magnetic regenerator. *Int. J. Refrigeration*, 32(6), 1478–1486

Pecharsky, V.K. and Gschneidner Jr., K.A. (1997). Giant magnetocaloric effect in Gd5(Si2Ge2). *Phys. Rev. Lett.*, 78(23), 4494–4497.

Petersen, T.F., Pryds, N., Smith, A., Hattel, J., Schmidt, H., Knudsen, H. (2008b). Two-dimensional mathematical model of a reciprocating room-temperature active magnetic regenerator. *Int. J. Refrigeration*, 31, 432–443.

Plait, A. (2019). Modélisation multiphysique des régénérateurs magnétocaloriques. PhD Thesis, Université Bourgogne Franche-Comté.

Roudaut, J. (2011). Modélisation et conception de systèmes de réfrogération magnétique autour de la température ambiante. PhD Thesis, Université de Grenoble.

Skokov, K.P., Karpenkov, D.Y., Kuz'min, M., Radulov, I., Gottschall, T., Kaeswurm, B., Fries, M., Gutfleisch, O. (2014). Heat exchangers made of polymer-bonded La(Fe,Si)13. *J. Appl. Phys.*, 115(17), 17A941.

Stevert, W.A. (1978). Stirling magnetic refrigerators and heat engines for use near room temperature. *J. Appl. Phys.*, 49, 1216–1226.

Tesla, N. (1889). Thermo-magnetic motor. US Patent US396121.

Trevizoli, P., Nakashima, A., Peixer, G., Barbosa Jr., J. (2017). Performance assessment of different porous matrix geometries for active magnetic regenerators. *Appl. Energy*, 187, 847–861.

Tura, A. and Rowe, A. (2011). Permanent magnet magnetic refrigerator design and experimental characterization. *Int. J. Refrigeration*, 34, 628–639.

Warburg, E. (1881). Magnetische Untersuchungen. *Ann. Phys.*, 249(5), 141–164.

Wu, J., Liu, C., Hou, P., Huang, Y., Ouyang, G., Chen, Y. (2014). Fluid choice and test standardization for magnetic regenerators operating at near room temperature. *Int. J. Refrigeration*, 37, 135–146.

Yao, G.H., Gong, M.Q., Wu, J.F. (2006). Experimental study on the performance of a room temperature magnetic refrigerator using permanent magnets. *Int. J. Refrigeration*, 29, 1267–1273.

You, Y., Wu, Z., Xiao, S., Li, H., Xu, X. (2017). A comprehensive two-dimensional numerical study on unsteady conjugate heat transfer in magnetic refrigerator with Gd plates. *Int. J. Refrigeration*, 79, 217–225.

Yu, B., Liu, M., Egolf, P.W., Kitanovski, A. (2010). A review of magnetic refrigerator and heat pump prototypes built before the year. *Int. J. Refrigeration*, 33, 1029–1060.

Zimm, C., Jastrab, A., Sternberg, A., Pecharsky, V.K., Gschneidner Jr., K.A., Osborne, M., Anderson, I. (1998). Description and performance of a near-room temperature magnetic refrigerator. *Adv. Cryog. Eng.*, 43, 1759–1766.

Zimm, C., Sternberg, A., Jastrab, A.G., Boeder, A.M., Lawton, L.M., Chell, J.J. (2001). Rotating bed magnetic refrigeration apparatus. US Patent No. 6526759.

6

Thermoelectric Systems as an Alternative to Reverse Cycle Engines

Julien RAMOUSSE[1] and Stéphane PAILHÈS[2]

[1] LOCIE, UMR 5271, CNRS, Université Savoie Mont Blanc, Le Bourget du Lac, France

[2] ILM, UMR 5306, UCBL, CNRS, Université de Lyon, Villeurbanne, France

Thermoelectricity (TE) today makes it possible to design devices in different industrial sectors such as micro/nano-electronics, automotive and aerospace for refrigeration and temperature control, design energy self-sufficient systems and also offers the opportunity for alternative energy by recycling waste heat into electricity. Its physical origin is based on three phenomena of coupled heat and mass transfers in condensed matter, discovered in the 19th century: the Seebeck effect, the Peltier effect and the Thomson effect (Goldsmid 2016). The scheme in Figure 6.1(a) gives an example of a commercial TE device with a wide choice of electrical characteristics that can be purchased for a few tens of euros. A TE module consists of alternating segments of "*n*"- and "*p*"-type semiconductor materials interconnected by metal electrodes and held between two insulating substrates (ceramics) acting as heat exchangers. Such a device can be used in **generator mode** when a temperature gradient is applied or in **heat pump mode** when it is traversed by an electrical

For a color version of all the figures in this chapter, see www.iste.co.uk/bonjour/refrigerators.zip.

Refrigerators, Heat Pumps and Reverse Cycle Engines, coordinated by Jocelyn BONJOUR.

current as shown schematically in Figures 6.1(b) and (c). The elementary technological building block of a TE module, called the "TE leg", is made up of a pair of "*p*"- and "*n*"-type semiconductor material segments. As shown later, an "*n*"-type semiconductor TE material generates an electromotive force under the effect of a temperature gradient (Seebeck effect) causing charges to move from the hot surface to the cold surface of the device. A "*p*"-type semiconductor TE material generates a charge flow in the opposite direction. Thus, a TE leg behaves like an electric generator when its opposite surfaces are maintained at different temperatures. Reversibly, it behaves like a heat pump when an electrical current passes through it. In this case, the heat is absorbed on one of its surfaces and released on the opposite surface, depending on the direction of the applied electrical current. The characteristics and performance of the complete module depend on its implementation (geometry of the legs, thermal and electrical contacts at the TE material/electrode/exchanger interfaces, etc.) and on the range and gradient of the temperature used. Since the 1990s, motivated by ecological and societal reasons, the funding of research in the field has made it possible to make remarkable advances both in the fundamental aspect to develop new concepts in order to enhance the efficiency of the thermoelectric conversion, making it possible to design new approaches to materials, and in the technological aspect, for the elaboration and characterization of nanostructured materials, and finally the production of devices covering a wide range of application sectors and temperature ranges, variable powers, specific needs or constraints on shaping, toxicity and cost. This chapter offers a current and pedagogical overview of thermoelectricity and is structured into two main parts oriented to the use in heat pump mode. *In the first part (section 6.1)*, the physical mechanisms behind thermoelectric effects are discussed with the aim of providing a basic understanding of the challenges and strategies of current research. *In the second part (section 6.2)*, the implementation and performance of TE systems in heat pump mode are discussed in order to identify the prospects for improvement according to their integration. A summary of the main applications is given at the end of the chapter in order to illustrate the wide field of application of this technology. References to recent specialized articles and books are offered to the reader in order to deepen knowledge on the topics discussed. The following books comprehensively cover the subject ranging from physical effects, development of materials and devices (Nolas et al. 2001; Lee 2016; Skipidarov and Nikitin 2016; Dávila et al. 2017; Rowe 2018).

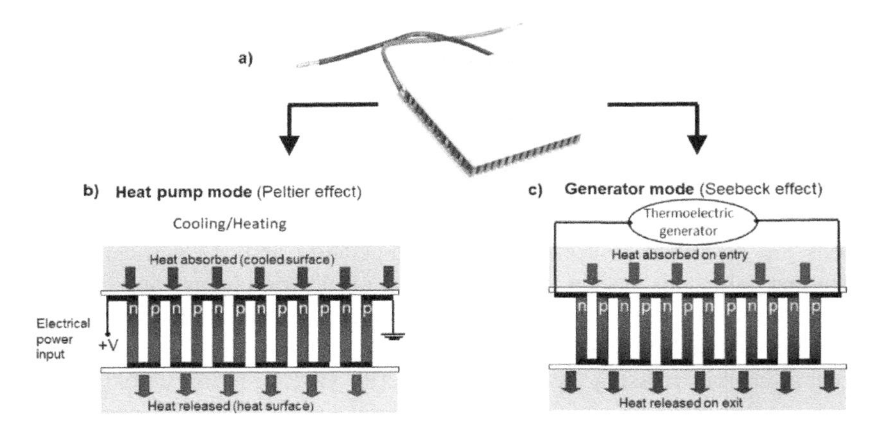

Figure 6.1. *The two ways of using a thermoelectric (TE) module*

NOTES ON FIGURE 6.1.– *a) Photograph of a TE module, Model ET-241-14-15 (P=105 W, 6 A, 29.8 V (dc)) from Adaptive Thermal Management, optimized for a temperature gradient of approximately 76°C with a dimension of 5 cm by 5 cm. Schematic representations of a cross-section of a TE module in b) heat pump and c) generator operating modes. A TE module is made up of "p"- and "n"-type semiconductor material segments that are interconnected in series in an alternating manner by metal electrodes (thick black lines) supported by ceramic insulating substrates (white rectangles). The color gradient indicates a temperature gradient (hot in red and cold in blue). b) Heat pump mode (refrigerator/heater): an electric power is applied to the input of the TE module, which is then crossed by an electrical current (Ohm's law). The exchanges of energy necessary for the passage of the charge current at the interfaces between the electrodes and the materials (Peltier effect) are responsible for the cooling of one of the surfaces of the module and the heating of the opposite surface. By changing the direction of the electrical current, the cooled and heated surfaces are reversed. b) Generator mode: a temperature gradient (ΔT) is imposed, the module is then crossed by a heat flux (Fourier's law), which generates an electromotive force thanks to the Seebeck effect (S). The module behaves as a generator characterized by an electromotive force $S.\Delta T$ and an internal resistance associated with the resistances of the material segments "p" and "n".*

6.1. Thermoelectricity fundamentals

In this section, we describe the physical microscopic mechanisms at the origin of the thermoelectric effects (TE), which appear within the material that loses thermodynamic equilibrium under the effect of an electrical current and/or a temperature gradient. The microscopic description is made at a level allowing understanding of the current research strategies. We limit ourselves to the system $\{R_1\text{-TE-}R_2\}$ formed from a single segment of TE material connected to two metal electrodes seen as charge and heat **reservoirs** and named R_1 and R_2 as shown in Figure 6.2. In this part, it is assumed that the **junctions** between the TE material and the reservoirs are thermally and electrically perfect. The effects of imperfect junctions are described in section 6.2. A temperature and/or electrical potential gradient applied to this system induces currents of electric charges and heat at the origin of the TE effects. In **heat pump mode** (Figure 6.2a), an electrical potential difference is applied between the two reservoirs (R_1 and R_2) which is at the origin of a current of charges crossing the whole system $\{R_1\text{-TE-}R_2\}$. In this case, the energy exchanges at the interfaces between the TE material and the reservoirs, necessary for the passage of the electrical current, are responsible for cooling or heating the surroundings closed to the contacts between the TE and the reservoirs. In **generator mode** (Figure 6.2b), the temperatures of the reservoirs are maintained at different values. The resulting temperature gradient induces a displacement of the charges, which can be used to power something like a bulb.

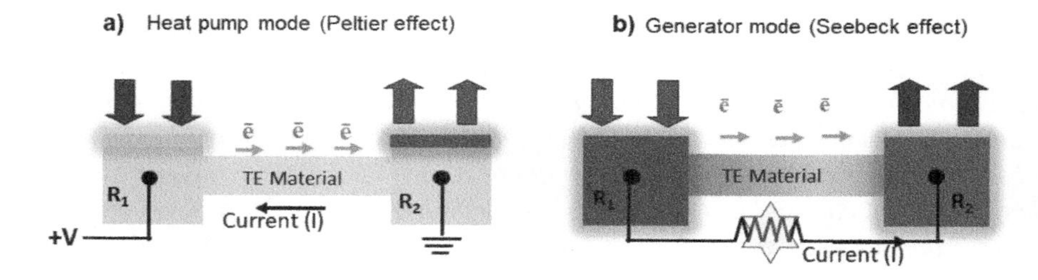

Figure 6.2. *Simple thermoelectric (TE) system {Reservoir(R_1)–TE Material–Reservoir (R_2)} in its two operating modes. The semiconductor "n"-type TE material is held between two reservoirs (R_1 and R_2)*

NOTES ON FIGURE 6.2.– *a) Heat pump mode (Peltier): the application of a potential difference (+V on R_1) generates an applied electrical current (I) which induces exchanges of energy necessary for the charges (ē) to cross the contacts between the TE material and $R_{1,2}$. These exchanges are at the origin of a heating and a cooling of the junctions and therefore of heat exchanges with the outside (symbolized by the blurred red and blue). b) Generator mode (Seebeck). The temperatures of the reservoirs are fixed at different values inducing the displacement of the charges (ē) and therefore the appearance of a current (I) when a charge is connected.*

In crystalline systems, at the microscopic scale, the TE effects are determined by the properties of the elementary excitations or quasi-particles associated with the collective motions of electrons and atoms, described by plane waves, and by the exchanges of energies at the reservoir interfaces, which is introduced in section 6.1.1. The expressions of the charge and heat currents, as well as the TE effects, are obtained from their spectral properties (wave vector, frequency or energy, polarization and lifetime) and from the statistics of these excitations. These microscopic formulations are given within the framework of Landauer's formalism in section 6.1.2. In section 6.1.3, the fundamental equations in thermoelectricity are given and the TE figure of merit or "ZT" is introduced, which allows us to gauge the TE efficiency of a material. The derivation of the ZT is made in section 6.2. In section 6.1.4, an overview is given of recent advances, known TE materials and major current avenues of research in materials for increasing ZT. The reader will find more complete microscopic descriptions of these effects in the following books: Datta (2005), Behnia (2015), Datta (2018) and Neophytou (2020).

6.1.1. *Transport of charge and heat*

6.1.1.1. *Concept of quasi-particles*

At the microscopic scale, crystalline matter is made up of billions of atoms (around 10^{22} atoms/cm^3) whose large-scale spatial ordering is obtained by repeating, by symmetry operations, the positions of a small group of atoms arranged in a so-called "elementary" unit cell whose characteristic size is in the range of one nanometer (Hammond 2015). Due to thermal agitation, the atoms held together by electrostatic forces vibrate around their equilibrium positions. According to their frequencies (ϑ), the collective vibrations of atoms are responsible for the propagation of sound ($\vartheta \sim$ kHz) and heat ($\vartheta \sim$Thz). The thermal energy also induces collective movements of the delocalized electrons, which are at the origin of the electrical current and also contribute to the transport of heat. Thus, heat and electrical current propagate mainly via discrete sets of elementary collective excitations, or eigenstates of excitations, of electronic or atomic origin. These elementary excitations are described by a set of plane waves each characterized by a propagation wave vector (k), a polarization (ε), an energy (E) and a dispersion relation linking the energy to the wave vector $E(k)$. The energy E is linked to the angular frequency ($\omega = 2\pi\vartheta$) by the Dirac constant: $E = \hbar\omega$. The wave group velocity ($v_g(k)$) is obtained from the local slope of $E(k)$: $v_g(k) = dE(k)/dk$ (Marder 2010). These eigenstates of excitations are obtained by solving the coupled equations of motion (Newton's law), a so-called "**semi-classical**" approach, or the Schrödinger equation, a "**quantum**" approach, for atoms and electrons. In a crystal, calculations are performed at the unit-cell level and are replicated by symmetry at infinity through the use of periodic boundary conditions. Defects or other interactions breaking the symmetry of the unit cell are generally dealt with in "**perturbation theory**". The knowledge of the eigenmodes of excitations and their statistics makes it possible to describe the macroscopic properties at equilibrium and out of thermodynamic equilibrium. In the particle view, these plane waves are associated with **quasi-particle states** (Marino 2017). The term "quasi" reminds us that it is about collective excitations of charges and atoms that are seen as individual particles. The properties of the quasi-particles carrying the charge can be very different from that of the isolated electron. The quasi-particles associated with the vibrations of atoms are called "phonons". Modern ab initio digital tools (Sholl and Steckel 2009) and spectroscopic experimental approaches such as angle-resolved photoemission spectroscopy (ARPES) (Hüfner 2003) or inelastic neutron scattering (INS) or inelastic X-ray scattering (IXS) (Pailhès et al. 2017) make it possible to map these states in a four-dimensional space, called a "**phase space**", containing an axis for the energy (*E*) and three axes for the wave vectors ($\vec{k}(k_x, k_y, k_z)$). $E(\vec{k})$ can form one or more complex geometrical shapes in this space. Examples of experimental maps in the phase space of electronic states in molybdenum disulfide (MoS_2) and phonons in germanium clathrate are shown in Figures 6.3(a) and (b) respectively. These are 2D cross-sections of $E(\vec{k})$ and $E(q)$[1] in the phase space in terms of energy and along certain directions in the reciprocal space. The points of strong intensities position the states of the quasi-particles or, in other words, the channels through which charge and heat are stored and transported in the material. Figures 6.3(c) and (d) show simplified representations of them in the case of a "*p*"-type semiconductor. The electronic structure is represented in terms of density of states $\rho_{el}(E)$, which corresponds to the integration in wave vectors of the states available at energy E+/-dE and shows a valence band (positive curvature) and a conduction band (negative curvature). The filling of its bands occurs up to an energy known as the "electrochemical potential" (μ). In a simplified diagram, $\rho_{el}(E)$ is represented by a valence band and a conduction band separated by a forbidden energy band or "**gap**", which does not contain a state. In the case of a "*p*"-type semiconductor (Figure 6.3(c) left), μ is close in energy (around k_BT) from the top of the valence band whereas in an "*n*"-type semiconductor, μ is located close to the conduction band. Only electronic states whose statistics can be thermally excited contribute to electronic transport. These are localized states in a small energy range centered on μ with an amplitude of a few k_BT. The value of μ, which varies little along the TE material, is controlled by doping and by the voltage difference applied

1. Traditionally, the wave vectors of the electron states are given by the letter "k", the letter "q" is used for those of the phonons.

in $R_{1,2}$ (Yu and Cardona 2010). The spectrum of phonon excitations is generally simplified by considering a single linear and isotropic dispersion (Debye regime), i.e. a $E(q) = c|q|$ with c being the average speed of sound, which disperses until it intersects with the energies of the optical phonon modes represented by a continuum of excitations, as shown in Figure 6.3(d). In general, the propagative contribution to heat transport by acoustic phonons is much greater than the diffusive contribution of optical phonons. Thermal conduction is therefore often described within the framework of the kinetic theory of gases where the free and independent particles are phonons in the Debye part of the spectrum (Lory et al. 2017).

As a first approximation, the charge and phonon transport equations are not coupled. In reality, the existence of a coupling can give rise to significant effects such as giant Seebecks of several tens of mV/K (Zhou et al. 2015). Thus, the two sets of quasi-particles can be associated with two gases of independent and indistinguishable quasi-particles whose transport is limited by diffusion processes of different natures. The macroscopic quantities at thermodynamic equilibrium are determined from the knowledge of the dispersion relations, therefore from $E(k)$ for the electrons and $E(q)$ for the phonons, and their occupancy statistics. Electronic and phononic states obey different filling statistics: the Fermi statistic for charge states ($f(E)$) and Bose–Einstein statistics ($n_B(E)$) for phonons. These two functions are fundamentally different, which gives these gases very different properties. Indeed, while the maximum value of the occupation of an electronic state is 1 (Pauli's exclusion principle), a phonon state can be thermally populated without limit. For phonons, there is therefore no equivalent to the electrochemical potential μ.

$$\begin{cases} f(E) = \frac{1}{1+\exp\left(\frac{E(k)-\mu}{k_B T}\right)} \\ n_B(E) = \frac{1}{exp\left(\frac{E(q)}{k_B T}\right)-1} \end{cases} \qquad [6.1]$$

The occupation of an electronic state ($E(k)$) or phononic state ($E(q)$) is directly related to the relationship between the difference $(E(k) - \mu)$ or $(E(q))$ and the thermal energy available ($k_B T$). Phononic (U_{ph}) and electronic (U_{el}) internal energy of the crystal is obtained by summing the energy of each state weighted by its occupation:

$$\begin{cases} U_{el} = \int Ef(E)\rho_{el}(E)dE \\ U_{ph} = \int En_B(E)\rho_{ph}(E)dE \end{cases} \qquad [6.2]$$

where $\rho_{el/ph}(E)$ correspond to the energy densities of the electronic (el) and phononic (ph) systems (i.e. the number of states contained in a section E+/-dE).

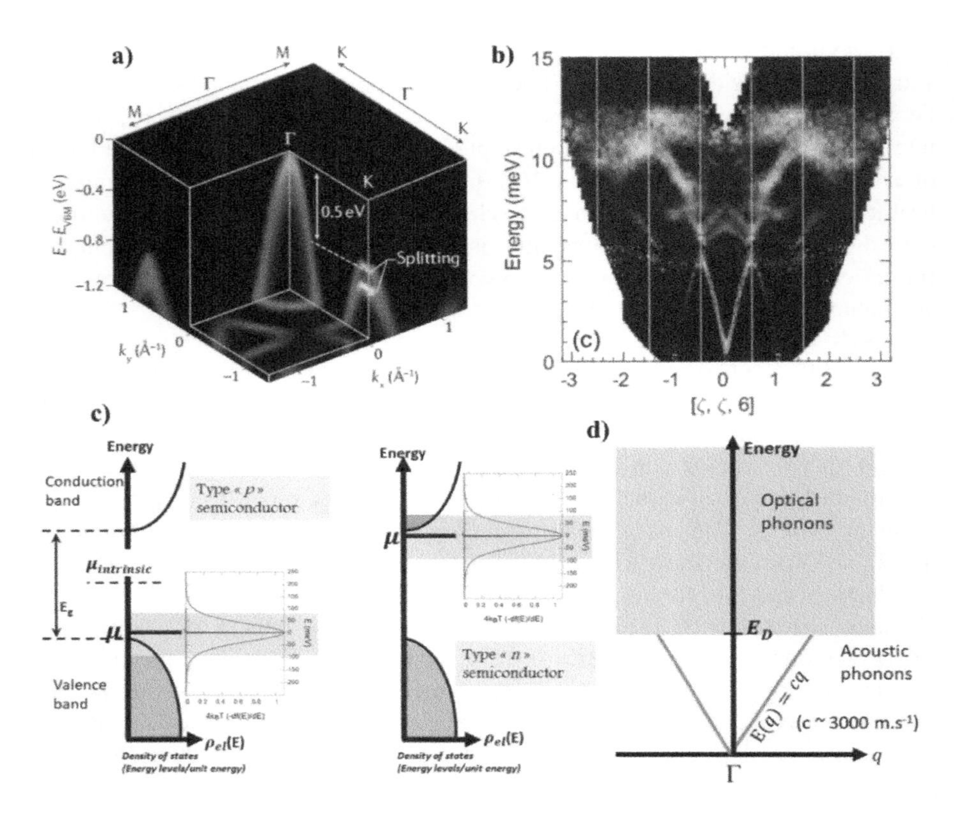

Figure 6.3. *Electronic and vibrational (phonons) quasi-particle states in phase space in wave vectors and in energy*

NOTES ON FIGURE 6.3.– *(a) Reprinted adapted with permission from Yuan et al. (2016). Copyright 2022 American Chemical Society. Map of the electronic bands, obtained by angle-resolved photoemission, in energy (measured with respect to the maximum of the valence band, E_{VBM}) and following two directions of wave vectors (k_x,k_y) in molybdenum disulfide MoS_2 (letters indicate points/directions of high symmetry). (b) Adapted from Turner et al. (2021). Slice of phonon dispersion, obtained by inelastic neutron scattering, in energy and along a wave vector direction in the material $Ba_{7.81}Ge_{40.67}Au_{5.33}$ of the clathrate family. The points of strong intensities (yellow-red) in the maps a) and b) localize the electronic and phononic states. (c) Simple representations of the electronic density of states as a function of energy ($\rho_{el}(E)$) in a semiconductor in the case where a valence band is separated from a conduction band by a so-called "gap" energy (E_g). When the electrochemical potential (μ) is close to the conduction band, the conduction is "n" type, and when it* *is close to the valence band, the conduction is "p" type. (d) Simple representation of a phonon spectrum including isotropic acoustic dispersion (blue, $E(q) = cq$, where c is the average speed of sound), and dispersing up to a maximum energy associated with a Debye energy (E_D) from which the spectrum is seen as a continuum of non-dispersive optical branches.*

6.1.1.2. *Contact with reservoirs*

In the system {Reservoir-Material TE-Reservoir}≡{R_1-TE-R_2} represented in Figure 6.2, the TE material is placed between two heat reservoirs (R_1 and R_2), thermally regulated at different temperatures, with which it can exchange charges and energy. In statistical physics, the thermal and chemical equilibrium of this set corresponds to that of a so-called "**grand canonical ensemble**" (Di Castro and Raimondi 2015). Simplified representations of the filling of electronic states in R_1, R_2 and the TE material are shown in Figure 6.4 in the heat pump operating mode (Figures 6.4(a)) and in the generator operating mode (Figure 6.4(b)). As discussed above and illustrated in Figure 6.3(c), the simplified band structure of the TE material schematized in Figures 6.4(a) and (b)) is representative of an "*n*"-type semiconductor with an electrochemical potential (μ) close to the conduction band (CB). Electron fillings in R_1 and R_2 are represented by colored bars indicating a filling up to the electrochemical potentials (μ_{R1} and μ_{R2}). The application of a positive voltage +V on R_1 implies that $\mu_{R1}>\mu_{R2}$. It is shown that the electrochemical potential μ in the TE material evolves linearly from μ_{R1} at the R_1-TE interface to μ_{R2} at the TE-R_2 interface. **Particle number conservation** implies that if a number N_{R_1-TE} of charges pass from R_1 to TE, a number of opposite charges N_{TE-R_2} is exchanged between TE and R_2:

$$N_{R_1-TE} + N_{TE-R_2} = 0 \Leftrightarrow N_{R_1-TE} = -N_{TE-R_2} = N \qquad [6.3]$$

Assuming that there is no energy dissipation (inelastic processes) inside the TE material, if the N charges which cross the R1-TE interface carry an energy E_{R_1-TE}, the **first law of thermodynamics** requires that a quantity of opposite energy E_{TE-R_2} must be carried by the charges passing through the TE-R_2 interface:

$$E_{R_1-TE} + E_{TE-R_2} = 0 \qquad [6.4]$$

The possible exchanges of energy and charges in the system {R_1-TE-R_2} are constrained by the **second law of thermodynamics**. The energy exchanges of the system {R_1-TE-R_2} with the environment are assumed to be only related to energy exchanges at the contacts between the TE material and the reservoirs $R_{1,2}$ (and therefore to the environment since R_1 and R_2 are thermally regulated). Note that inelastic processes existing in the channel, not considered here, can be modeled by the addition of a third reservoir, the latter can be the phonon bath as an example. The reservoirs $R_{1,2}$ are metal electrodes each characterized by an electrochemical potential $\mu_{R1/R2}$ and a temperature $T_{R1/R2}$. By definition, these reservoirs have a very large number of degenerate electronic configurations to distribute the charges, and which is much higher than that of the TE material. Being metallic, the dispersion relation of the electronic states of a reservoir is quadratic, $E(k) \propto k^2$ (Fermi liquid), leading to a density of electronic states, which increases by the square root of the energy, $\rho_{R_{1,2}}(E) = \partial E_k/\partial k \propto \sqrt{E}$. Thus, with regard to the electronic states available at a given energy, **it is statistically more favorable to transfer energy to the reservoir than to take energy from it**. This fundamental property of the reservoir, which dictates the direction of the energy exchanges in contact with the TE material, is linked to the notion of entropy and to the second law of thermodynamics. Indeed, a major contribution of Boltzmann's microscopic approach is the link between the thermodynamic definition of entropy, $\frac{1}{T} = \left(\frac{\partial S}{\partial E}\right)$ (where T is the temperature), and the available microscopic electronic configurations. In a given energy range (E +/- dE), the latter is given by the density of states $\rho_{el}(E)dE$ (Yoshioka 2007, Guthmann et al. 2007):

$$S(E) = k_B \ln(\rho_{el}(E).dE) \quad [6.5]$$

Consider a functioning of the system {R_1-TE-R_2} in the heat pump mode (Peltier effect) where under the effect of a voltage applied between R_1 and R_2, which changes the relative positions of μ_{R1} and μ_{R2}, an isothermal current of charges appears. In the case of the situation illustrated in Figure 6.4(a), the transfer of N charges of energy $N\mu_{R1}$ in R_1 at a higher energy corresponding to unoccupied states in the conduction band of the TE material requires an average energy that will be noted "ε". This energy is recovered from the environment, and the interface is then cooled. If the charges flow in the opposite direction, the interface is heated. The statistics associated with the transfer of the N charges of R_1 to the TE material are determined by the ratio of the numbers of electronic configurations available in R_1 at energies $N\mu_{R_1}$ and $N\mu_{R_1} + \varepsilon$. Since $\rho_{R_{1,2}}(E)$ is an increasing function with E, the statistic favors the case where the energy increases in the reservoir, so $\varepsilon \geq 0$. This results in a condition on the entropy variation of R_1. Assuming $\varepsilon \ll N\mu_{R_1}$, the ratio of probabilities is written as (Datta 2005):

$$\frac{\rho_{R_1}(N\mu_{R_1}+\varepsilon)}{\rho_{R_1}(N\mu_{R_1})} \sim e^{\frac{\varepsilon}{k_B}\left(\frac{\partial S}{\partial E}\right)} = e^{\frac{\Delta E_{R_1}}{k_B T_{R_1}}} \geq 1 \Leftrightarrow \Delta S_{R_1} \geq 0 \quad [6.6]$$

Similar reasoning applies to the TE-R_2 interface. Thus, the entropy balance of the system {R_1-TE-R_2} is: $\Delta S_{R_1} + \Delta S_{R_2} \geq 0$. If the system {$R_1$-TE-$R_2$} is isothermal, the inequality is verified only if the global energy variation of R_1 and R_2 is positive which is only possible if the system {R_1-TE-R_2} receives energy. There are sources of entropy variation within the TE material such as internal microscopic processes of inelastic diffusion. This can be accommodated by adding other reservoirs in contact with the TE material and will not be developed here.

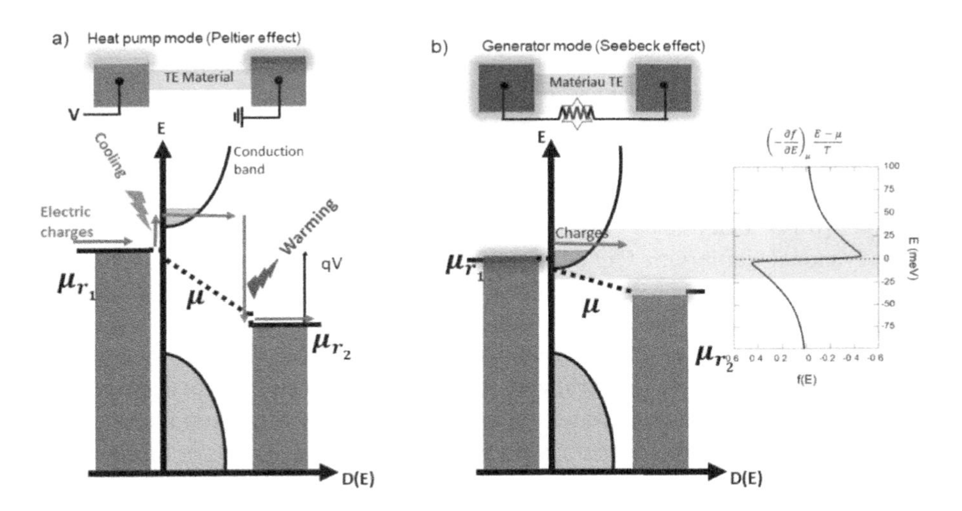

Figure 6.4. *Charge and energy transfer phenomena in the system {Reservoir (R_1)–type-n thermoelectric material (TE)–Reservoir (R_2)}. The orange vertical rectangles represent the electronic filling in R_1 and R_2 up to electrochemical potentials μ_{R1} and μ_{R2} respectively. The simplified band structure of the TE material is "n" type (see Figure 6.3(c))*

NOTES ON FIGURE 6.4.– *(a) Heat pump mode: a voltage applied between R_1 and R_2 changes the relative positions of μ_{r1} and μ_{r2} and is at the origin of an isothermal electronic current flowing from R_1 to R_2. This current is associated with a cooling of the contact between R_1-TE and a warming between TE-R_2. (b) Generator mode: a temperature gradient is imposed between R_1 and R_2, here $T_{R1}>T_{R2}$, the imbalance of the occupancy of the electronic states in the TE material induces a charge movement from hot (red) to cold (blue) and the appearance of a voltage ($\mu_{r1}\neq \mu_{r2}$). The induced current is calculated by integrating the electronic states in an energy window given by the antisymmetric product $\left(-\frac{\partial f_0}{\partial E}\right)\times\left(\frac{E-\mu}{qT}\right)$ (blue) and represented in the inset (see text).*

6.1.2. *Thermoelectric effects*

Thermoelectric (TE) effects in the {Reservoir-TE Material-Reservoir} system≡{R_1-TE-R_2} correspond to physical effects out of thermodynamic equilibrium associated with the appearance of a charge current I, and heat current I_Q, under the effects of a temperature gradient ΔT and/or electrical potential ΔV. The application of a ΔV induces a variation in the electrochemical potential within the TE material ($\Delta\mu$) between values μ_{R_1} in R_1 and μ_{R_2} in R_2 such that $\Delta\mu = \mu_{R_1} - \mu_{R_2} = e\Delta V$. As explained previously, the two sets of eigenmodes of excitations of charges and atoms are described by two gases of independent and indistinguishable elementary quasi-particles whose knowledge of their equilibrium statistics, see equation [6.1], determines the macroscopic properties of the material at thermodynamic equilibrium. Out of thermodynamic equilibrium, under the effects of external disturbances such as ΔV and/or ΔT, the statistical distributions change leading to variations at the macroscopic scale of so-called "intensive" thermodynamic variables (electrochemical potential, temperature). Within the formalism of **Onsager's generalized theory of irreversible processes**, the gradients $\{\Delta T, \Delta V\}$ are associated with generalized forces which are at the origin of the currents $\{I_Q, I\}$ that describe the system response. Assuming a slow, or **quasi-static**, evolution of the system in front of the characteristic relaxation times of the two gases (~10^{-15} and ~10^{-12} s for the electron and phonon gases respectively), the non-equilibrium thermodynamic quantities are obtained by deriving the relations between the forces and the currents, which adjust to minimize the production of entropy. This constitutes the main hypothesis of Onsager's principle, which assumes that any fluctuating intensive variable undergoes a restoring force towards equilibrium by coupling effect with the others, which is found in the Le Chatelier–Braun principle. By limiting this to linear terms (linear response theory), the currents of heat and charges $\{I_Q, I\}$ are related to the forces $\left\{\frac{1}{T}\Delta\mu, \Delta\frac{1}{T}\right\}$ via the system of equations (Apertet et al. 2013):

$$\begin{cases} I = eL_{11}\frac{1}{T}\Delta\mu + eL_{12}\Delta\frac{1}{T} = \frac{e^2L_{11}}{T}\Delta V - \frac{eL_{12}}{T^2}\Delta T \\ I_Q = L_{21}\frac{1}{T}\Delta\mu + L_{22}\Delta\frac{1}{T} = \frac{eL_{21}}{T}\Delta V - \frac{eL_{22}}{T^2}\Delta T \end{cases} \quad [6.7]$$

Transport coefficients L_{ij}, also called "**kinetic coefficients**", are intimately related to the microscopic properties of the two quasi-particle gases. If the non-equilibrium processes described in Onsager's principle are irreversible, the symmetry is not affected by time reversal. This means that the principle of micro-reversibility applies and implies the following relationship: $L_{21} = L_{12}$. The reader will find detailed information on the Onsager principle in Goupil (2016). In the following, the charge and heat currents are described microscopically and linked to the kinetic coefficients L_{ij}.

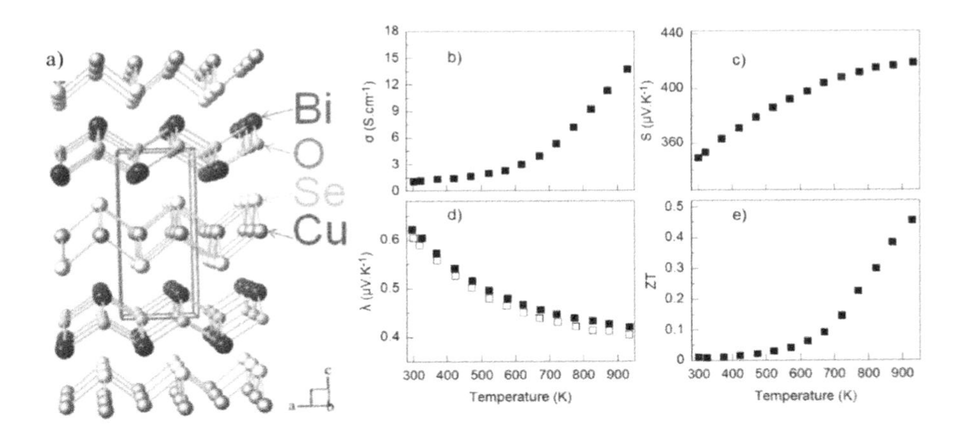

Figure 6.5. *Adapted from L.-D. Zhao et al. (2014). Examples of thermoelectric (TE) temperature measurements in the oxychalcogenide BiCuSeO TE material. a) Crystallographic structure consisting of an alternation of Bi_2O_2 and Cu_2Se_2 planes. Temperature dependences between 300 and 900 K of b) electrical conductivity (σ), c) Seebeck coefficient (S), d) thermal conductivity (λ) total (solid squares) and lattice (open squares) and e) of the figure of merit TE (ZT)*

6.1.2.1. *Isothermal electronic conductivity*

An electrical potential difference is applied between R_1 and R_2 while the system $\{R_1\text{-TE-}R_2\}$ is maintained at the same temperature, *T*. The first term in equation [6.7] comes down to Ohm's law:

$$[I]_{\Delta T=0} = eL_{11}\frac{1}{T}\Delta\mu = e^2L_{11}\frac{1}{T}\Delta V \equiv G_0\Delta V \qquad [6.8]$$

where $G_0 = \frac{e^2L_{11}}{T}$ is the electronic conductance related to the electronic conductivity (σ) and the electrical resistivity (ρ = 1/σ) by a so-called "**geometric**" factor given by the ratio between the length (L) of the TE material and its cross-section (S): $G_0 = \sigma\, L/S$. At the microscopic level, the application of a ΔV between R_1 and R_2 modifies the electrochemical potential across the TE material, which induces a current of charges. For an energy level *E*, the latter is written as:

$$I(E) = G(E)\left[f_{R_1}(E,\mu_{R_1},T) - f_{R_2}(E,\mu_{R_2},T)\right] \qquad [6.9]$$

where *G(E)* is the conductance between the two electrodes associated with a quasi-particle state of energy E in the TE material and $f_{R_1/R_2}(E)$ corresponds to the values of the Fermi distribution taken at the level of the R_1-TE and TE-R_2 interfaces respectively. The function *G(E)* depends on the density of electronic states and the mean free path at energy E. Following Landauer's principle, the charge is assumed to travel through the material while remaining on the same energy level, and the path is said to be "elastic" (Di Ventra 2010). In the limit where the applied voltage is low compared to the thermal energy ($k_B T$~25 meV at 300 K), the above difference can be linearized and, using the properties of partial derivatives, it is shown that:

$$I(E) = G(E)\left(-\frac{\partial f}{\partial E}\right)_{\mu} \Delta V \quad [6.10]$$

Thus, the electrical conduction of an energy level E is linked to the conductance of the levels located in an energy window of $4k_B T$ given by the derivative of the Fermi statistic in μ. This integration window is represented in Figure 6.3(c) in both cases "*p*"-type and "*n*"-type electronic band structure of the electronic band structure. The total electrical current is obtained by integrating equation [6.10] on the energy. Therefore:

$$I = \left[\int_{-\infty}^{+\infty} dE G(E)\left(-\frac{\partial f}{\partial E}\right)_{\mu=\mu_0}\right] \Delta V \equiv G_0 \Delta V \quad [6.11]$$

Figure 6.5(b) shows an example of the temperature evolution above ambient temperature of the electrical conductivity (σ) in the thermoelectric compound BiCuSeO. The range of characteristic values in TE materials is a few tens of $S.cm^{-1}$. Since TE materials are generally semiconductors, σ decreases by lowering the temperature which simply results from the closing of the integration window ($4k_B T$) and the absence of electronic states available around μ.

6.1.2.2. *The Seebeck effect*

The Seebeck effect (*S*) is associated with the electrical voltage induced under the effect of a temperature gradient when the electrical circuit is open. If I = 0 in the first term of equation [6.7], then: $\frac{e^2 L_{11}}{T}[\Delta V]_{\text{open circuit}} - \frac{e L_{12}}{T^2}\Delta T = 0$. By definition:

$$S \equiv \frac{[\Delta V]_{\text{open circuit}}}{\Delta T} = \frac{L_{12}}{e T L_{11}} \quad [6.12]$$

Thus, under the effect of a temperature gradient, the TE material behaves like a voltage generator characterized by an electromotive force $S\Delta T$ and an internal resistance given by G_0^{-1}. Microscopically, the temperature gradient unbalances the occupations of the electronic states through the TE material which induces a charge current and therefore an electrical potential difference which, in turn, modifies the electrochemical potentials in R1 and R2. In open circuit and, after reaching steady state, the two effects compensate each other and the resulting voltage is the Seebeck voltage. For an energy level E, the charge current is written as:

$$I(E) = G(E)[f(E, \mu_1, T_1) - f(E, \mu_2, T_2)] \quad [6.13]$$

The linearization of this difference allows the following to be written as:

$$I(E) = G(E)\left(-\frac{\partial f_0}{\partial E}\right) \times \left(\frac{1}{q}\Delta\mu + \frac{E-\mu}{qT}\Delta\mathrm{T}\right) \quad [6.14]$$

Assuming elastic trajectories of the charges, the total charge current is obtained by integrating equation [6.14]:

$$I = \underbrace{\left[\int_{-\infty}^{+\infty} dE\left(-\frac{\partial f}{\partial E}\right)_{\mu} G(E)\right]}_{G_0} \mathrm{e}\Delta\mathrm{V} + \underbrace{\left[\int_{-\infty}^{+\infty} dE\left(-\frac{\partial f}{\partial E}\right)_{\mu} \frac{E-\mu}{qT} G(E)\right]}_{G_S} \Delta\mathrm{T} \quad [6.15]$$

where μ is the electrochemical potential in the material. The charge current breaks down into two terms, the first corresponds to the ohmic conductance (G_0) previously described, the second reflects the conductance associated with the temperature difference (G_S). The latter is antisymmetric with respect to μ and can be negative or positive implying that the associated charge current can move from the hot electrode to the cold electrode or vice versa. The sign depends on the electronic band structure of the material depending on the product $(E - \mu)G(E)$. This integration window is represented in Figure 6.4(b). If the density of electronic states in the TE material is an increasing function of energy (type "n" conduction), there are more thermally activated states with energies greater than μ. In this case, the charges move from the hot reservoir to the cold reservoir. Conversely, if the density of states is a decreasing function of energy ("p"-type conduction), there are more thermally depopulated states with energies below μ. In this case, the charges move from the cold reservoir to the hot reservoir. The Seebeck voltage is the open-circuit potential difference. The total charge current is then zero (I = 0), and equation [6.15] gives:

$$G_0\,[\Delta V]_{\text{open circuit}} + G_S\,\Delta T = 0 \quad [6.16]$$

The Seebeck coefficient is defined by:

$$S \equiv \frac{[\Delta V]_{\text{open circuit}}}{\Delta T} = -\frac{G_S}{G_0} \quad [6.17]$$

As in the case of an "n"-type band, $G_S > 0$ and $S < 0$. Conversely, in the case of a "p"-type band, $G_S < 0$ and therefore $S > 0$. Figure 6.5(c) shows an example of high temperature evolution of the Seebeck coefficient in the compound BiCuSeO. The range of characteristic values in TE materials is a few hundred μV/K. The Seebeck coefficient decreases in temperature, which is directly related to the reduction in temperature of the integration window $(E - \mu)G(E)$ described above.

6.1.2.3. *The Peltier effect*

The Peltier effect corresponds to the heat flow carried by a current of isothermal charges when the two reservoirs R_1 and R_2 have the same temperature (T). In this case, an electrical potential difference $\Delta\mu = e\Delta V$ is applied between R_1 and R_2 by an external generator. The heat current passing through the material is:

$$[I_Q]_{\Delta T=0} = L_{21}\frac{1}{T}\Delta\mu = \frac{L_{21}}{eL_{11}}[I]_{\Delta T=0} \equiv \Pi[I]_{\Delta T=0} \quad [6.18]$$

where Π defines the Peltier coefficient. Microscopically, when a charge passes through the TE material by loaning an energy level E, quantities of energy are exchanged in contact with R_1 and R_2 to allow the charge to pass from the energy E in the TE material to the last occupied electronic levels R_1 and R_2 with energies μ_{R1} and μ_{R2}. The interfaces are either heated or cooled depending on the relative positions of E with respect to μ_{R1} and μ_{R2}. A Peltier operating mode of system {R_1-TE-R_2}, where the charge current is responsible for cooling R_1 and warming R_2 in the case of an "n"-type TE material, is represented in Figure 6.4(a). For an energy level E, the heat current resulting from the energy transfers at the R_1-TE and TE-R_2 interfaces is written as:

$$I_Q(E) = G(E)\left[f(E,\mu_{R_1},T_{R_1})(E-\mu_{R_1}) - f(E,\mu_{R_2},T_{R_2})(E-\mu_{R_2})\right] \quad [6.19]$$

The linearization of this expression and the integration on the energy make it possible to obtain the expression of the total heat current according to the potential difference and the temperature gradient applied between R_1 and R_2:

$$I_Q = \underbrace{\left[\int_{-\infty}^{+\infty} dE\, G(E)\left(-\frac{\partial f_0}{\partial E}\right)\frac{(E-\mu_0)}{q}\right]}_{G_P}\Delta V + \underbrace{\left[\int_{-\infty}^{+\infty} dE\left(-\frac{\partial f}{\partial E}\right)_{\mu=\mu_0}\frac{(E-\mu_0)^2}{q^2T_0}G(E)\right]}_{G_Q}\Delta T \quad [6.20]$$

The Peltier heat current is associated with the heat current induced by an isothermal charge current *($\Delta T = 0$)* passing through the TE material and transferred to the reservoirs. The Peltier coefficient is defined by the ratio of charge and heat currents:

$$\Pi \equiv -\frac{G_P}{G_0} \qquad [6.21]$$

As the expressions of G_p and G_S indicate, the Peltier and Seebeck coefficients are linked by $\Pi=TS$. This relationship, known as the first Kelvin relation, is a direct consequence of the micro-reversibility of the irreversible processes involved (Apertet and Goupil 2016).

6.1.2.4. *The Thomson effect*

The Thompson effect appears when a non-isothermal charge current passes through the TE material. In this case, a current induced by the Seebeck effect appears in addition to the ohmic current and is at the origin of energy exchanges by Peltier effect between neighboring segments of the TE material. The microscopic description of the Thompson effect will not be discussed here.

6.1.2.5. *Thermal conduction*

6.1.2.5.1. Thermal conduction of electronic origin

We have previously expressed the heat current (I_Q) as a function of electronic thermal conduction G_Q. This is the thermal conduction of the system in a short circuit, i.e. for $\Delta V = 0$ but $\Delta T \neq 0$, therefore without transfer of charge. A heat current of electronic origin can exist without being associated with a charge current by compensation of the charges but not of the energy. In this case, the energies transported by the charges going from one reservoir to another are different and the overall energy current is not zero. This open-circuit thermal conduction is directly related to the kinetic coefficient L_{22} via relation: $\left[I_Q\right]_{\Delta V=0} = -\frac{L_{22}}{T^2}\Delta T \equiv -G_Q \Delta T$. The electronic thermal conduction, G_K, or open-circuit thermal conduction, corresponds to the flow of heat which passes through the material at zero charge current, $I = 0$, under the effect of a temperature gradient. By imposing the condition $I=0$ in the system of equations [6.7], where:

$$\left[j_Q\right]_{j=0} = -\left[\frac{L_{11}L_{22}-L_{12}^2}{T^2 L_{11}}\right]\Delta T \equiv -G_K \Delta T \qquad [6.22]$$

The first term in G_K corresponds to the transport of heat without charge transfer, therefore to G_Q. The second is associated with the transfer of charges and can be related to the Seebeck, Peltier and electrical conductance coefficients such that:

$$G_K = G_Q - S\Pi G_0 = G_Q - \frac{G_S G_P}{G_0} \qquad [6.23]$$

6.1.2.5.2. Lattice thermal conductivity

In most TE materials, the thermal conduction is dominated by the crystal lattice contribution and not by the electronic contribution. As discussed above, the set of crystal lattice excitations can be seen as a gas of independent quasi-elementary particles (phonons) described by Bose–Einstein statistics; see equation [6.1]. The energy flow associated with the contribution of an energy level E is in the approximation of an "elastic" path:

$$I_{Q,Ph}(E) = G_{ph}(E)\,[n(E,T_1) - n(E,T_2)] \times E \qquad [6.24]$$

where $G_{ph}(E)$ is the thermal conduction associated with the energy level E, which depends on the vibrational density of states and the mean free path of the phonons at the energy level E. By linearizing the difference around the mean temperature T and by summing over the set of available energy levels, the heat flux associated with lattice vibrations is:

$$I_{Q,Ph} = \left[\underbrace{\frac{k_B^2 T}{\hbar}\int_0^{+\infty} dE\, G_{ph}(E)\left(\frac{E}{k_B T}\right)^2\left(\frac{\partial n}{\partial E}\right)}_{G_{K,Ph}}\right]\Delta T \equiv G_{K,Ph}\Delta T \qquad [6.25]$$

where $G_{K,Ph}$ is the lattice thermal conduction which is related to the lattice thermal conductivity (λ_{Ph}) by the geometric factor $G_{Ph} = \lambda_{Ph}\, {}^{A}/_{L}$. The total thermal conductivity of the material (λ) is obtained by the sum of two contributions:

$$\lambda(T) = \lambda_e(T) + \lambda_{Ph}(T) \qquad [6.26]$$

where λ_e is the thermal conductivity of electronic origin. Figure 6.5(d) shows an example of the temperature evolutions of $\lambda(T)$ and $\lambda_l(T)$ in the compound BiCuSeO. The range of characteristic values of λ in TE materials is around the $W.m^{-1}.K^{-1}$ and is dominated by its phononic component. The temperature behavior of λ_{Ph} is not trivial to understand. At very low temperatures, λ_{Ph} *(T)* tends to zero, which is directly related to the Bose–Einstein statistic. By heating, *lambda*$_{Ph}$ *(T)* passes through a temperature maximum around 50 K (associated with the so-called "Umklapp" peak) above which it decreases continuously due to the activation of inelastic scattering processes between phonons. The reader will find a detailed description of the behavior of λ_{Ph} *(T)* in the following works: Tritt (2004) and Kaviany (2008).

6.1.2.5.3. The Wiedemann–Franz law

The Wiedemann–Franz law relates thermal conductivity of electronic origin (λ_e) to the isothermal electrical conductivity (σ) by a linear relationship characterized by a factor "L_0" called the "**Lorentz factor**". It describes the transport of energy associated with charge transfer:

$$L_0 = \frac{\kappa_e}{T\sigma} = \frac{L_{11}L_{22} - L_{12}^2}{e^2 T^2 L_{11}^2} \qquad [6.27]$$

This law implies that a good electrical conductor is also a good thermal conductor since the electronic component of the overall thermal conductivity of the material is proportional to its electrical conductivity. Once the Lorentz factor has been determined, it is possible to subtract the electronic contribution from the thermal conductivity measurement to obtain the lattice contribution.

6.1.2.6. *Fundamental equations and figure of merit*

The TE coefficients were expressed as a function of transport integrals ($G_{0,S,p,Q}$) on the microscopic states of electronic and phononic elementary quasi-particles. The kinetic coefficients L_{ij} of the system of equations [6.7] describing charge and heat currents under the effects of temperature and electrical potential gradients are described at the microscopic level by transport integrals ($G_{0,S,p,Q}$) calculated from the spectra of electronic quasi-particles and phonons and knowledge of their mean free paths (Datta 2005):

$$\begin{cases} I = G_0\,\Delta V + G_s\,\Delta T \\ I_Q = G_p \Delta V + G_Q \Delta T \end{cases} \qquad [6.28]$$

with:

$$\begin{cases} G_0 = \int_{-\infty}^{+\infty} dE \left(-\frac{\partial f}{\partial E}\right)_{\mu=\mu_0} G(E) \\ G_s = \int_{-\infty}^{+\infty} dE \left(-\frac{\partial f}{\partial E}\right)_{\mu=\mu_0} \frac{E-\mu_0}{qT_0} G(E) \\ G_P = \int_{-\infty}^{+\infty} dE \left(-\frac{\partial f}{\partial E}\right)_{\mu=\mu_0} \frac{E-\mu_0}{q} G(E) \\ G_Q = \int_{-\infty}^{+\infty} dE \left(-\frac{\partial f}{\partial E}\right)_{\mu=\mu_0} \frac{(E-\mu_0)^2}{q^2 T_0} G(E) \end{cases} \qquad [6.29]$$

The system of equations can be rewritten so as to reveal the physical observables, namely the electrical resistance (*R*), the Seebeck coefficient (*S*), the Peltier coefficient (*Π*) and thermal conduction (G_K):

$$\begin{cases} \Delta V = R\,I + S\,\Delta T \\ I_Q = -\Pi\,I + G_K \Delta T \end{cases} \quad [6.30]$$

with:

$$\begin{cases} R = \frac{1}{G_0} = \frac{T}{e^2 L_{11}} \\ S = -\frac{G_S}{G_0} = \frac{L_{12}}{eTL_{11}} \\ \Pi = -\frac{G_P}{G_0} = \frac{L_{21}}{eL_{11}} \\ G_K = \left(G_Q - \frac{G_P G_S}{G_0}\right) = \left[\frac{L_{11}L_{22} - L_{12}^2}{T^2 L_{11}}\right] \end{cases} \quad [6.31]$$

These fundamental equations make it possible to obtain the heat fluxes associated with the passage of an electrical current in a TE material subjected to a temperature gradient. From this, we determine the TE efficiency in generator mode and the coefficient of performance (for heating or cooling) in heat pump mode for a TE material, then for a couple of "*p*" and "*n*" materials, taking into account the geometry (section and length) of the materials. The expressions are then extended to the module by integrating the effects of the thermal and electrical resistances at the interfaces with the reservoirs. This is detailed in section 6.2. At this level, we retain that the efficiency of a TE material in both operating modes depends on an adimensional parameter intrinsic to the material known as the **figure of merit "ZT"** and equal to the product of the temperature (T) by a coefficient $Z = \frac{G_0 S^2}{K}$. Therefore:

$$ZT = \frac{G_0 S^2 T}{K} \quad [6.32]$$

The objective of the research is to identify physical concepts for producing new materials where this ZT factor is as large as possible. The temperature evolution of the ZT in the BiCuSeO compound is shown in Figure 6.5(e). Unfortunately, ZT augmentation presents major fundamental difficulties and significant advances will result from the emergence of new concepts. The first challenge is optimizing the numerator, also called the TE "**power factor**", $G_0 S^2$. Indeed, while the electrical conductivity increases with the charge carrier density, the Seebeck coefficient decreases and reaches its lowest values in metals. Thus, the power factor optimum for carrier densities is typically in the range 10^{18}–10^{20} cm^{-3}, which corresponds to the doping range of heavily doped semiconductors. Similarly, for the denominator, the electronic contribution to the thermal conductivity increases with the charge carrier density and, in metals, even becomes predominant compared to that of the lattice. These conflicting evolutions of the TE coefficients as a function of charge carrier density can be simply shown in the case of an electronic band structure with a single parabolic band, and in the case of the degenerate approximation, assuming the absence of strong electronic correlations and assuming energy-independent scattering processes. Simple expressions are obtained for the electronic conductivity, the Seebeck coefficient and the thermal conductivity of electronic origin in order to obtain the evolution of the ZT as a function of the charge carrier density in the material. The evolutions of G_0, S and Ke as a function of the density of charge carriers (n) are summarized in Figure 6.6. Detailed derivations are given in the following books and reviews: Behnia (2015) and Goldsmid (2016).

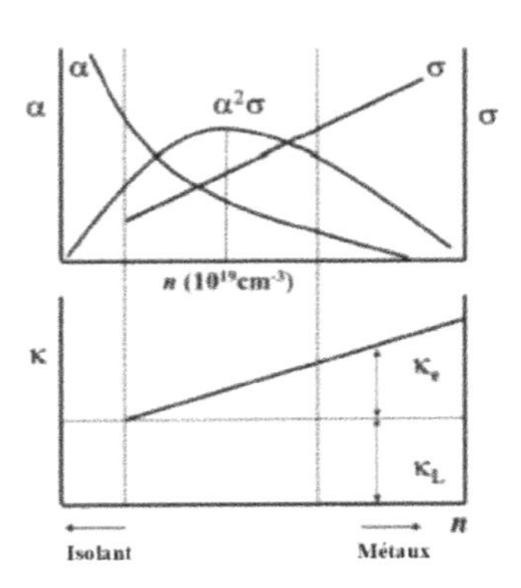

Figure 6.6. *Qualitative evolutions of the Seebeck coefficient (α), electrical conductivity (σ), thermoelectric power factor ($\alpha^2\sigma$), thermal conductivity of electronic origin (κ_e) and lattice thermal conductivity (κ_L)*

6.1.3. *Main lines of research*

6.1.3.1. *Overview of current performance*

Although TE effects have been known for almost two centuries, the development of TE materials only really began in the 1950s and 1960s after the discoveries of bismuth telluride Bi_2Te_3, lead telluride PbTe and Si-Ge alloys with figures of merit close to 1 and the first developments of thermoelectric devices for aerospace. After that and until the 1990s, the field experienced a period without much development. From the 1990s, a set of social, economic and scientific factors were behind the renewed interest and allowed the discovery of new families of materials with ZT exceeding 1. The need to deploy alternative energy sources to fossil fuels has made it possible to finance projects in the field. Scientifically, significant advances have been made both in the understanding of the physical phenomena of electronic and thermal transport behind the appearance of new concepts, some of which are linked to the nano revolution, the development of high-performance characterization tools both in the micro-structural aspects and on the characterization of the physical properties necessary to evaluate the figure of merit. Since then, research in the field has continued to evolve, particularly through the development of new numerical tools, leading to the development of new concepts guiding new material approaches and the multiplication of advanced methods for the development of nanostructured materials. It has also diversified both in the nature of the materials (inorganic, organic) and in the shaping (thin layers, composites) of the materials. The following journal articles report on these historical developments: Chen et al. (2003), Dresselhaus et al. (2007), Snyder and Toberer (2008), Minnich et al. (2009) and Polozine et al. (2014), Zhang and Zhao (2015). The worldwide effort to increase the material figure of merit is summarized in Figure 6.7. In parallel with the development of materials, applications have diversified into microelectronics, the power supply of autonomous devices, sensors, etc. (Bell 2008; Champier 2017; He et al. 2018; Jaziri et al. 2020).

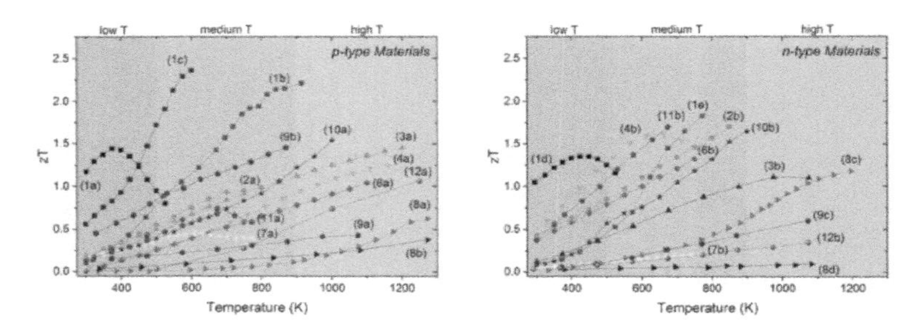

Figure 6.7. *Thermoelectric figure of merit (zT) as a function of temperature for the families of materials (Adapted from Freer et al.2022)*

NOTES ON FIGURE 6.7.– *(1) tellurides: p-type—(1a) Bi0.5Sb1.5Te3, (1b) Pb0.98Na0.02Te—4%SrTe, (1c) Ge0.86Pb0.1Bi0.04Te, n-type—(1d) Bi1.8Sb0.2Te2.7Se0.3, (1e) PbTe—4%InSb; (2) skutterudites: p-type—(2a) CeFe3.85Mn0.15Sb12, n-type—(2b) Ba0.08La0.05Yb0.04 Co4Sb12; (3) half Heuslers: p-type—(3a) Nb0.88Hf0.12FeSb, n-type—(3b) Zr0.2Hf0.8NiSn0.985Sb0.015; (4) Zintls (including Mg3Sb2): p-type—(4a) Yb14Mn0.2Al0.8Sb11, n-type—(4b) Mg3Sb1.5Bi0.5; (6) clathrates: p-type—(6a) Ba8Ga15.8Cu0.033Sn30.1, n-type—(6b) Ba8Ga16.6Ge28.7; (7) FeGa3-type materials: p-type—(7a) RuGa2.95Zn0.05, n-type—(7b) FeGa2.80Ge0.20; (8) actinides and lanthanides: p-type—(8a) Yb3.8Sm0.2Sb3, (8b) USi3, n-type—(8c) La3Te4, (8d) URu2Si2; (9) oxides: p-type—(9a) Ca2.8Bi0.2 Co4O9, (9b) Bi0.94Pb0.06Cu0.99Fe0.01SeO, n-type—(9c) Sr0.95(Ti0.8Nb0.2)0.95Ni0.05O3; (10) sulfides and selenides: p-type—(10a) Cu2Se, n-type—(10b)*

Pb0.93Sb0.05S0.5Se0.5; (11) silicides: p-type—(11a) Mg2Li0.25Si0.4Sn0.6, n-type—(11b) Mg1.98Cr0.02(Si0.3Sn0.7)0.98Bi0.02; (12) Borides and Carbides: p-type—(12a) Boron carbide (13.3 at.% C), n-type—(12b) Ca0.5Sr0.5B6.

Around room temperature, the most effective inorganic compounds are alloys derived from bismuth telluride (Bi_2Te_3), lead telluride (PbTe), antimony telluride (Sb_2Te_3), silicon-germanium alloy (SiGe) and tin selenide (SnSe). The different strategies for optimizing the thermoelectric performance of materials are detailed in the following general reviews: Shi et al. (2016), Yang et al. (2016) and Ma et al. (2021).

6.1.3.2. *Nanostructuration*

The nanostructuring of solid materials, i.e. with macroscopic sizes, offers the possibility of impacting thermal transport with lesser effects on electronic transport. This is possible due to differences in the mean free paths associated with the dominant wavelengths of electronic quasi-particles for electrical conduction and phonons for lattice thermal conductivity. The increase in interface density on a scale around a hundred nanometers affects the transport of phonons much more than that of charges. On the other hand, with the dominant wavelengths as well as the nature of the diffusion processes being dependent on the nature of the material and the temperature, nanostructuring is only effective in a certain temperature range. The objective is therefore to tune the characteristics of the nanostructures to target the spectral range of the phonons which dominate the thermal transport by impacting as little as possible the electronic conduction. Nanostructuring is achieved by different material approaches which can be combined by acting either at the microstructure scale, at the grain scale or on the grain boundaries. Thus, the design and realization of nanostructures for thermoelectricity constitutes a vast field of current research. The reader can find detailed information with recent references in the following journal articles: He et al. (2013), Eibl et al. (2015), Fitriani et al. (2016), Jiang et al. (2016), Alam and Ramakrishna (2017) and Hao et al. (2019).

The nanostructuring of the microstructure of solid materials is obtained by the rapid consolidation of powders known as "Spark Plasma Sintering" (SPS). Thanks to the use of a pulsed current and high speeds of temperature rise, this method makes it possible to significantly reduce the duration of the densification process, thus limiting granular growth. This results in a dense material with grain sizes of hundreds of nanometers. Flash sintering is now commonly used in the shaping or development (in the case of reactive sintering) of TE materials which significantly improves the "ZT". TE temperature measurements between ambient and 700 K obtained for the compound $In_{0.25}Co_4Sb_{12}$, nanostructured by SPS, varying the average grain size are shown in Figure 6.8. Reducing the average grain size reduces the lattice thermal conductivity (λ_l) in this material with greater efficiency around room temperature (see Figure 6.8b). The temperature evolution of the figure of merit (ZT) for the different average grain sizes shows an optimum (see Figure 6.8c). This indicates that the increase in ZT linked to the reduction in lattice thermal conductivity no longer compensates for the reduction in the TE power factor.

The introduction of a microstructure at the scale of nano-grains makes it possible to study the impact of nanostructuring in solid materials over a size range of a few nanometers, much lower than that of nano-grains. The granular nanostructuring, which often appears naturally, can be achieved and controlled by metallurgical approaches consisting of heat treatments and involving phase transformations or precipitations. There are several possible morphologies of intra-granular nanostructures: nano-inclusions, lamellar microstructures or modulated structures. If from a fundamental point of view, nanostructuring makes it possible to improve the TE efficiency of materials, it is important that it be stable under the conditions of use of the devices (Sharma and Sugar 2014).

The development of TE nano-materials (thin films, nano-wires, membranes, etc.) has been motivated by technological demands in micro-electronics and by general concepts on the amplification of the Seebeck coefficient by quantum confinement effects in 1D or 2D structures, which induces singularities of the electronic density of states at Fermi levels and on the differences of the mean free paths of phonons and charge carriers. Thus, significant increases in the Seebeck coefficient and ZT have been achieved in thin layers, superstructures or nanowires of conventional TE materials such as Bi_2Te_3, PbTe, etc. The reader can find more detailed information in the following journal articles: Chen et al. (2003), Dresselhaus et al. (2007) and Mao et al. (2016b).

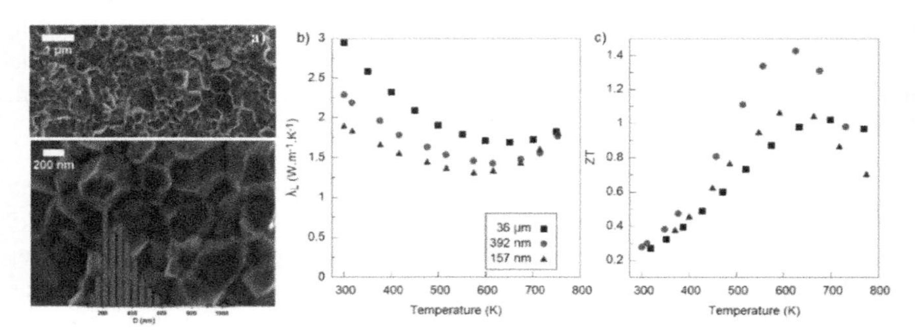

Figure 6.8. *Adapted from Benyahia et al. (2018). (a) Electron microscopy images of a surface obtained after fracture of a solid nanostructured sample of the material $In_{0.25}Co_4Sb_{12}$ at two magnifications allowing the scales indicated in the figures to be reached. The average distribution of grain diameters (D) is represented by the red bar graph. (b) Lattice thermal conductivity (λ_L) and (c) thermoelectric figure of merit (ZT) measured for three different average grain sizes in the compound $In_{0.25}Co_4Sb_{12}$*

6.1.3.3. *Structural complexity*

The structural complexity is associated with the complexity of the atomic organization in the crystal and chemical bonding inhomogeneity at the scale of the elementary crystallographic unit, therefore at a sub-nanometric scale. Two examples of complex crystallographic meshes corresponding to a so-called "cage" material from the family of inorganic clathrates and a complex material from the family of tetrahedrites are represented in Figure 6.9. The design of complex crystallographic lattices is one of the main approaches to increase the ZT (Toberer et al. 2010) (Snyder and Toberer 2008). Indeed, this makes it possible to decouple and therefore to act independently on thermal transport and electronic transport (Toberer et al. 2010, 2011). By definition, a complex lattice has a large number of atoms distributed over crystallographic sites of different point symmetry (Wyckoff positions). The demultiplication of these sites opens up numerous possibilities of substitutions making it possible to vary the electrical transport properties and the Seebeck coefficient without affecting the thermal conductivity too much. It is thus possible, in the same material and by finely adjusting the composition, to obtain a semiconductor whose dominant conduction character is either of the "*n*" or "*p*" type. But the most remarkable effect is certainly the very significant reduction in thermal conductivity down to values below the $W.m^{-1}K^{-1}$ and which is associated with the complexity of the crystallographic lattice, i.e. large number of atoms and chemical bonding inhomogeneity. Indeed, increasing the structural complexity in a crystal makes it possible to reduce the propagative transport of heat in favor of a diffusive mode of transport (Pailhès et al. 2014; Turner et al. 2021). Thus, a germanium or silicon clathrate composed of more than 80% germanium/silicon has a thermal conductivity at room temperature 2 orders of magnitude lower than that of Ge/Si diamond. The difference increases at low temperatures up to 3 orders of magnitude. In the case of the clathrate lattice made of cages, this has been interpreted as the effect resulting from the reduction of the phase space in energy and in momentum available for the acoustic phonon states (Euchner et al. 2012, Pailhès et al. 2014). The lattice of the tetrahedrites does not present any cages, but their thermal conductivity is just as low as that of the clathrates. This is the consequence of a second effect, induced by the complexity, which creates strong anharmonicity (Bouyrie et al. 2015).

In crystalline matter, organic crystals give an upper limit to structural complexity. Organic conductors such as conjugated polymers or so-called coordination polymers, made up of metal centers, suspended in an organic matrix via complex ligands, offer new perspectives for applications and also for improving the ZT. Even if the current state of the art of the ZT of these materials remains well below the values obtained in inorganic ones, the immense diversity of the achievable structures makes it possible to vary the complexity, the dimensionality of the electronic conduction channels or the porosity, and may allow important advances in the years to come. They have the advantage of having intrinsically very low lattice thermal conductivity, typically around 0.5 $W.m^{-1}.K^{-1}$, due to the organic matrix which is responsible for heat diffusion. Several recent review articles report the state of the art of thermoelectricity in these materials (Leong et al. 2016; Cowen et al. 2017; Wang et al. 2021).

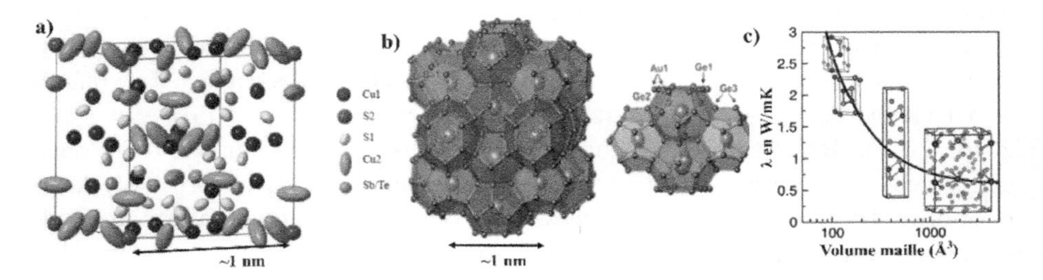

Figure 6.9. *Complex crystallographic meshes at the nanometric scale*

NOTES ON FIGURE 6.9.– *a) Adapted from Bouyrie et al. (2015). Representation of the cubic crystallographic structure of the compound Cu12Sb2Te2S13 from the tetrahedrite family. The colors of the atoms on the different crystallographic sites are indicated in the inset legend. The atomic volumes (ellipsoids) are representative of the atomic thermal agitation factors obtained by X-ray diffraction measurements. b) Adapted from Lory et al. (2017). Representation of the cubic crystallographic structure of the compound* $Ba_{7.81}Ge_{40.67}Au_{5.33}$ *of the clathrate family. It consists of a lattice of two types of germanium (Ge) cages, cages with 20 (dodecahedron) and 24 (tetradecahedron) Ge atoms, which encapsulate barium atoms. The substitution of certain Ge atoms, here by gold atoms (Au), makes it possible to modify the electronic and thermal properties. c) Adapted from Toberer et al. (2010), Lattice thermal conductivity (λ) at 300 K represented as a function of the volume of the crystallographic lattice in a series of materials called "Zintl" antimonides.*

6.1.3.4. *Electronic effects*

Different approaches have been developed to amplify the ZT figure of merit by increasing the numerator G_0S^2, known as the TE power factor (see equation [6.32]). These involve acting on the electronic band structure or on the nature of the electronic diffusion processes and make it possible to increase the Seebeck coefficient with little impact on the electrical conductivity. In the following sections, the main concepts are discussed, such as band convergence, resonant states, spin-orbit coupling and energy filtering. The following reviews give more detailed explanations: Mehdizadeh Dehkordi et al. (2015), Yang et al. (2016), Ma et al. (2021).

It is advantageous to have several electronic bands of the same concavity ("*n*" or "*p*" type) close to the Fermi level (Pei et al. 2011). The overall contribution of these bands can be described by considering an equivalent electrical diagram of a parallel circuit, each branch of which is associated with a band. Thus, while the electronic conductivity increases due to the additive contributions of the carriers on the different bands, each band creates an electromotive force and therefore contributes to the overall Seebeck coefficient of the material. In general, this increases the power factor and is presumed to be one of the effects responsible for the high figures of merit in many families of thermoelectric alloys such as the PbTe chalcogenides (Jood et al. 2020), Mg_2(Si,Sn) (Liu et al. 2012; Mao et al. 2016a; Pandel et al. 2021), half-Heuslers (Zhu et al. 2019), etc. Nevertheless, to reach the maximum power factor of a multi-band electronic structure, it is necessary to align the energies of the different bands in an interval around the available thermal energy, i.e. k_BT (Kyu Hyoung Lee et al. 2020). This convergence of the bands can be achieved by adjusting the chemical composition, by changing the temperature or by applying mechanical stresses (Balout et al. 2017). The electronic structure at the Fermi level of the compound PbTe is composed of two families of "pockets" (marked by the letters "L" and "Σ") isolated from two different valence bands. The energy difference between these two bands in PbTe is 0.2 eV (more than 2,000 K) at room temperature, but is reduced very quickly by heating because the "L" band undergoes a strong temperature renormalization, while the "Σ" band is almost insensitive to it. Convergence is reached at a temperature of 450 K, which causes the rapid increase in the figure of merit. This convergence temperature of the bands can be reduced by the substitution effect of Te by Se, which tends to reduce the difference in energy between the two bands.

The Seebeck coefficient, which depends on the energy derivative of the electronic density of states, can be significantly amplified in the case of a rapid variation in energy of the electronic density of states close to the Fermi level. This is observed with the introduction of so-called "resonant" electronic levels (for a review, see Heremans et al. (2012)). These are electronic levels with very low energy dispersion that couple to electronic conduction states. The introduction of such levels can be achieved either by doping or by the presence of point defects in very dilute regimes. Co-doping is usually required to align the Fermi level to that of the resonant state. The amplification effect of the Seebeck coefficient associated with the effect of resonant levels has been observed in different families of thermoelectric materials such as lightly doped chalcogenides such as Tl-PbTe, Al-PbSe, In-SnTe (Misra et al. 2020) or even In-GeTe (Wu et al. 2017) and its theoretical study (Wiendlocha 2016; Thébaud et al. 2017). Another approach to produce singularities in the density of states close to the Fermi level is to artificially reduce the dimensionality of the band structure. This is observed in the case of solid materials without an inversion center of symmetry and with a very strong spin-orbit coupling, which has the effect of lifting the degeneracy of the two states of the spin of the electron (up and down) fluid friction. Thus, an electronic band appears for up-spins and another for down-spins whose minima are shifted in energy and in wave vector. This results in toroidal shapes of the Fermi surfaces whose density of state resembles that observed in the case of a two-dimensional system, which results in an amplification of the Seebeck coefficient. Such an amplification effect of the Seebeck effect associated with spin-orbit coupling has been observed in the family of compounds BiTeX (X=I, Br and Cl) (Maaß et al. 2016; Li et al. 2020) and lead telluride PbTe (Hong et al. 2020).

Finally, the energy filtering of the charge carriers is based on the difference of the transport integrals for the conductivity and the Seebeck coefficient, and the fact that the electronic states contributing in a predominant way to the electronic conductivity and to the Seebeck coefficient are distributed differently in the electronic band structure. Thus, in the case of an "n"-type semiconductor, for example, the suppression of the electronic states at the bottom of the band, the closest to the electrochemical potential with the smallest velocities, makes it possible to increase the Seebeck coefficient with less impact on electronic conductivity. A low-pass filter of these states makes it possible to block their conduction and has the consequence of amplifying the Seebeck coefficient. Such a filter can be implemented by introducing grain boundaries, the design and experimental implementation of which is a current issue because, in practice, the reduction in the associated electronic conduction remains the dominant effect (Narducci 2017; Gayner and Amouyal 2020; Lin et al. 2020).

6.2. Implementation and performance analysis

The TE effects described above make it possible to design devices that can either generate electrical power based on the Seebeck effect – **generator mode**, or transfer heat from a low-temperature heat reservoir to a high-temperature heat sink, based on the Peltier effect – **heat pump mode**. For this last operating mode, two distinct systems can be identified according to the intended useful effect (heating a high-temperature heat reservoir from low-temperature heat) – **heat pump for heating**, or cooling a low-temperature heat reservoir with high-temperature heat release) – **heat pump for cooling/refrigerator**. In the case of very specific applications where both effects are beneficial (simultaneous heating at high temperature and cooling at low temperature), the system will then be qualified as **double-function heat pump**.

6.2.1. *Implementation of thermoelectric modules*

A TE module is made of an association of elementary units called "thermoelectric legs" connected electrically in series and thermally in parallel. The faces are usually covered with ceramics which act as a heat exchanger and serve as mechanical support and electrical insulation between the TE legs. A TE leg is made up of one segment of an "*n*"-type and another "*p*"-type semiconductor materials that are connected by a metal electrode. When powered up, current flows from the "*n*"-type segment to a "*p*"-type segment through the metallic connection. The entire module constitutes an electrical device. Several architectures have been developed. The most classic for TE macro- and micro-modules is the vertical (or "π") architecture where the legs are arranged vertically with respect to the substrate as illustrated in Figure 6.10. The second widespread architecture, particularly for thin-film micro-modules on a rigid or flexible substrate, is the so-called planar architecture where the TE legs are printed or deposited on the substrate.

In heat pump mode, when a current flows through the device in one direction, one of the surfaces is cooled while the other is heated. If the current is reversed, the cooled and heated surfaces are also reversed. In generator mode, by applying a temperature gradient between the two surfaces, the module generates a voltage to supply a charge. Thus, a TE module is a solid system that can be used as desired for heating, cooling or electrical generation. These devices work without light or removable parts, limiting the generation of noise, and offer great stability.

Many factors must be taken into account for the design of a TE module depending on the targeted use (generator or heat pump) and the conditions of use. First, TE materials should be selected based on their performance in the targeted temperature range and taking into account their abundance, stability, ease of shaping and the possibility of having a "*p*" and "*n*" material pair with good performance in a close temperature range. If the temperature gradients are large, the temperature dependencies of the physical properties and of the TE efficiency of the materials must be taken into account. To optimize performance over the entire temperature range, it is possible to consider so-called "**multi-stage**" leg segments, i.e. they are made up of several materials or of the same material but with a concentration gradient of charge carriers (Ouyang and Li 2016; Yusuf and Ballikaya 2020;

El Oualid et al. 2021). The second step is the optimization of the geometry of the TE module as detailed below. The main optimization parameters are the leg length, the ratio of the surfaces between the "*n*"- and "*p*"-type segments and the density of segments. Minimizing the thermal and electrical resistances is also essential in order not to deteriorate the TE performance. Ceramics made of Al oxide (Al_2O_3) or nitride (AlN or Si_3N_4) must be very flat and polished (roughness < μm). Thermal pastes or a deposition of graphite layers are used to make up for contact faults at the interfaces. Particular attention must be paid to the material and the method used to assemble the TE materials to the metal electrodes (copper, aluminum or nickel). The aim is to limit the thermal stresses, to avoid the diffusion of matter, which could modify the properties of the TE materials, and to reduce the electrical resistances as much as possible. Depending on the temperature range, either lead-free solder is used for low temperatures, or aluminum or silver solders.

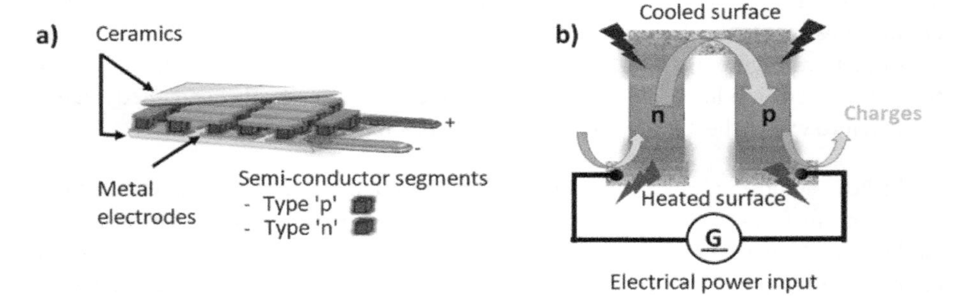

Figure 6.10. *Representation of a TE module in the so-called "vertical" architecture as shown in the scheme of Figure 6.1 a) The segments of the "p" (blue pads)- and "n" (red pads)-type TE materials are interconnected by metal electrodes, supported by ceramic insulating substrates. b) Illustration of the heat pump operating mode of a TE leg made up of a "p" and "n" segment and powered electrically. The isothermal electrical current flow induces cooling and heating of the opposite surfaces due to energy transfers at the TE material/metal electrode junctions (Peltier effect)*

6.2.2. *Performance analysis of thermoelectric modules*

The heat pump operation and the energy flows exchanged by the system is illustrated in Figure 6.11.

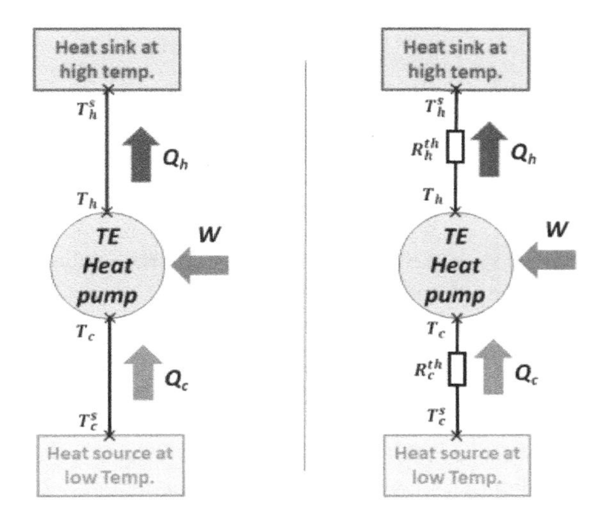

Figure 6.11. *Schematic representation of the energy flows exchanged between the thermoelectric system and its heat reservoirs (cold and hot) in heat pump mode. a) Ideal thermal couplings and b) with thermal resistances ($R_{c,h}^{th}$)*

The efficiency of heat pump systems is characterized by the coefficient of performance (COP), defined as the ratio of useful thermal power to the electrical power delivered to the system. It is therefore necessary to distinguish the COP_h in heating mode from COP_c in cooling mode, as follows:

– Heating mode:

$$COP_h = \frac{Q_h}{W} \qquad [6.33]$$

– Cooling mode:

$$COP_c = \frac{Q_c}{W} \qquad [6.34]$$

It should be noted here that by virtue of the energy conservation ($Q_h = W + Q_c$), these two definitions are linked by the following relation: $COP_h = COP_c + 1$. In order to introduce a conversion efficiency bounded between 0 and 1, it is proposed to use the following definition of the efficiency of a heat pump, independently of the desired useful effect:

$$\eta_{HP} = \frac{Q_c}{Q_h} = \frac{COP_c}{COP_h} \qquad [6.35]$$

6.2.3. *Intrinsic performance of thermoelectric systems*

From the description of the TE effects presented previously (section 6.1.2), heat flows Q_h and Q_c exchanged with the hot heat sink (*h*) or the cold heat source (*c*) by a single TE leg, characterized by an internal electrical resistance R_i, a Seebeck coefficient *S* and a Peltier coefficient Π and a thermal conduction *K* and flowed by an electrical current *I* are given by the sum of a heat flux exchanged at the junctions by the Peltier effect equal to $\Pi I = SIT_{h/c}$, a heat flux dissipated by the Joule effect distributed equitably between the hot and cold heat reservoirs given by $R_i I^2/2$ and a heat flux related to Fourier's law given by $K\Delta T$ (Apertet et al. 2013). For an entire module made up of *N* legs, the thermal (Q_h and Q_c) and electrical (*W*) powers exchanged are written as (Ioffe et al. 1959):

$$W = \Delta V . I = N(R_i I^2 + SI\Delta T) \qquad [6.36]$$

$$Q_h = N\left(SIT_h + {R_i I^2}/{2} - K\Delta T\right) \qquad [6.37]$$

$$Q_c = N\left(SIT_c - {R_i I^2}/{2} - K\Delta T\right) \qquad [6.38]$$

with R_i = |rho L_l / S_l

and K = |λ S_l / L_l

In this model, the physical properties of the material, which depend on the temperature, are taken at the average of the temperatures of the hot and cold reservoirs $T_m = \frac{T_h+T_c}{2}$. This approximation remains acceptable as long as the temperature difference between the reservoirs remains small (Yamashita 2008). This modeling was compared to different approaches in order to validate its accuracy, despite the simplicity of its implementation (Fraisse et al. 2013).

From equations [6.36]–[6.38], it clearly appears that the powers involved and the performance of the system depend on the electrical current supplied to the thermoelectric module. For illustration, the cold power Q_c and conversion efficiency

η_{HP} are represented in Figure 6.12, in the case of a Bi_2Te_3 module, for different electrical currents and junction temperature ratios.

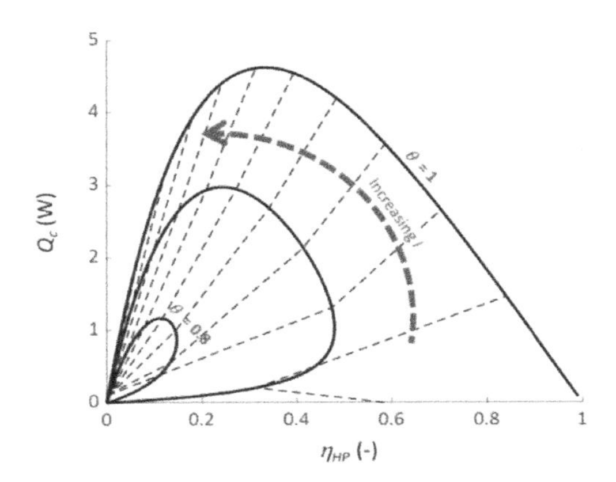

Figure 6.12. *Cooling capacity Q_c and conversion efficiency η_{HP} of a Bi_2Te_3 thermoelectric module (S = 230 10^{-6} V.K^{-1}; λ = 1.6 W.m^{-1}.K^{-1}; ρ = 1.4 10^{-5} Ω.m; L = 5 10^{-3} m; A = 1 10^{-6} m^2), for electrical currents varying between 0 and 1.6 A (in increments of 0.2 A) and junction temperature ratios between 0.8 and 1 (in increments of 0.1), with a constant average temperature of 350 K. The solid lines are drawn for constant junction temperature ratios, and the dotted lines for constant electrical currents*

The influence of the temperature ratio of the heat reservoirs associated with the thermoelectric module is in line with the theory of reversible heat pumps, known as Carnot heat pumps: the lower the temperature difference between the two heat reservoirs (here junction temperatures), the better the conversion performance of the Heat Pump system. As a reminder, the Carnot coefficients of performance of reversible heat pump systems are given by:

– Heating mode:

$$COP^C{}_h = \frac{T_h}{\Delta T} \qquad [6.39]$$

– Cooling mode:

$$COP^C{}_c = \frac{T_c}{\Delta T} \qquad [6.40]$$

That is, for the expression of the efficiency used here:

$$\eta_{HP}^{C} = \frac{COP^{C}{}_{c}}{COP^{C}{}_{h}} = \frac{T_c}{T_h} = \theta \qquad [6.41]$$

As a result, when the temperature ratio of the heat reservoirs decreases, the actual conversion performance, as well as the thermal powers, of the system decrease (see Figure 6.12). In addition, it should be noted here that the operation of reversible systems in heat pump mode is conditioned by the junction temperatures and the TE properties of the material used. In order to obtain a cooling power ($Q_c > 0$), it is necessary to respect the following condition:

$$\frac{{T_c}^2}{\Delta T} > \frac{2}{Z} \qquad [6.42]$$

where ZT is the figure of merit TE introduced in section 6.1.1.5. The electrical current range for which operation in heat pump mode is achieved is then limited by:

$$\frac{S}{R}\left(T_c - \sqrt{T_c^2 - \frac{2\Delta T}{Z}}\right) < I < \frac{S}{R}\left(T_c + \sqrt{T_c^2 - \frac{2\Delta T}{Z}}\right) \qquad [6.43]$$

The electrical current applied to the TE module (and therefore the corresponding voltage $\Delta V = R\,I + S\,\Delta T$) determines the conversion efficiency and the thermal powers exchanged. The operating conditions at maximum power are different from those at maximum efficiency (Figure 6.12). It is therefore necessary to define the objective aimed when implementing a thermoelectric module, depending on the intended application:

– **maximum thermal powers exchanged**, if priority is given to the useful effect (heating or cooling), to the detriment of the system efficiency. This strategy will be favored in cases where strong operational constraints predominate over operating costs;

– **maximum conversion efficiency**, if priority is given to the efficient use of the thermal energies exchanged, to the detriment of the thermal powers exchanged. This strategy will be preferred in the case of limited thermal resources (in terms of availability and/or cost);

– in general, the question is to define the **trade-off between thermal power exchanged and conversion efficiency**, depending on the intended application, and the associated technical and economic constraints.

The optimum operating conditions (at maximum power and at maximum efficiency) are summarized in Table 6.1 (Ramousse and Goupil 2018).

Operating condition	$Max(\eta_{HP})$	$Max(Q_c)$
I (A)	$\frac{K\Delta T}{ST_m}(1+M)$	$\frac{ZKT_c}{S}$
Q_c (W)	$\frac{NK\Delta T}{T_m}\left[(M+1)\left(T_c - \frac{\Delta T}{2(M-1)}\right) - T_m\right]$	$NK\Delta T\left(\frac{ZT_c{}^2}{2\Delta T} - 1\right)$
η_{HP} (–)	$\frac{\theta M - 1}{M - \theta}$	$\frac{ZT_c^2 - 2\Delta T}{ZT_c(2T_h + T_c) - 2\Delta T}$

Table 6.1. *Optimal operating conditions (maximum thermal power and maximum conversion efficiency) of a thermoelectric module (with $M = \sqrt{1+ZT}$)*

Finally, it should be noted that for any electrical current outside the optimal range (i.e. $\frac{ZKT_f}{S} \leq I \leq \frac{K\Delta T}{ST_m}(1+M)$), an adjustment of the electrical current applied makes it possible to improve the conversion efficiency and/or the thermal powers exchanged.

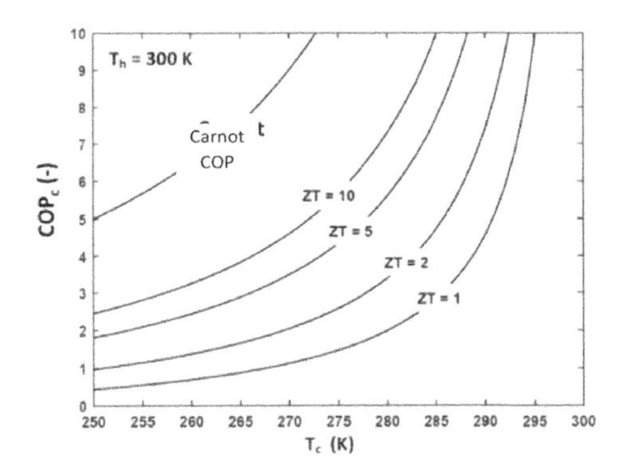

Figure 6.13. *Optimal cooling COP values of thermoelectric systems, as a function of material figure of merit ZT and cold junction temperature T_c for a hot junction temperature of T_h = 300 K*

With recent developments of new TE materials exhibiting higher figures of merit, the expected performance as a function of figure of merit ZT and cold junction temperature T_c are represented in Figure 6.13, for a hot junction temperature at T_h = 300 K. On the same curve, for a given ZT value, the COPs (hot and cold) decrease with the temperature gradient ΔT, in accordance with the theory of reversible heat pumps. Passing from curve to curve, for a fixed value of ΔT, the COP increases with ZT. Within the limit where $ZT \rightarrow \infty$, the COP tends towards the

Carnot COP. There is a priori no fundamental limit to the values of ZT, but, in practice, the current state of the art makes it possible to reach ZT values of around 1.5, corresponding to theoretical cooling COPs of 1 for a temperature gradient of 40 K.

6.2.4. *Optimal module design*

Based on the optimal operating conditions at maximum efficiency (leading to maximum COP) defined in Table 6.1, it is possible to propose a system sizing and control strategy to meet a given thermal demand (Ramousse et al. 2015). Hot and cold junction temperatures (T_h and T_c) and thermoelectric properties (S, λ, R and therefore ZT) are assumed to be known. The objective here is to determine the optimal geometric characteristics (number of thermoelectric legs N^*, leg length L_l^* and leg cross section S_l^*), as well as the corresponding optimal electrical current I^*, to meet a given thermal demand with optimum performance (maximum COP). The equations previously introduced in Table 6.1 show that any design that respects equations [6.44] and [6.45] will guarantee the satisfaction of the thermal need with maximum performance:

$$\frac{I^*.L_l^*}{S_l^*} = \frac{\lambda \Delta T}{S T_m}(1 + M) \quad [6.44]$$

$$\frac{L_l^*}{S_l^*.N^*} = \frac{\lambda \Delta T}{Q_u T_m}\left[(M + 1).\left(T_u + \frac{\Delta T}{2(M-1)}\right) - T_m\right] \quad [6.45]$$

where T_u and Q_u are associated with the desired useful effect ($T_u = T_h$ and $Q_u = Q_h$ in heating mode; or $T_u = T_c$ and $Q_u = Q_c$ in cooling mode). These conditions result in infinite possible designs that meet the objective (Khire et al. 2005; Kong Hoon Lee and Kim 2007; Pan et al. 2007). In order to converge towards a unique design, it is proposed to take into account the following additional criteria:

– Minimize the volume of thermoelectric material ($V = N^*.S_l^*.L_l^*$) used, in order to reduce raw material costs. Together with the previous equalities, this objective involves reducing the length of the thermoelectric legs.

– Reduce the number of thermoelectric junctions in order to limit the technical issues previously mentioned when implementing the thermoelectric material in a module. This objective results in the maximization of the leg cross-section and therefore of the electrical current applied to the module.

In conclusion, the classic design of TE modules, consisting of a large number of TE legs, with small cross-sections and (relatively) large length, can be revisited in the case where the compactness constraints are low, as proposed in Figure 6.14.

Furthermore, this proposal leads to the reduction of the thermal power densities exchanged, leading to reduced thermal resistance between the TE module and the heat reservoirs, thus improving the overall performance of the system, as discussed in the following section.

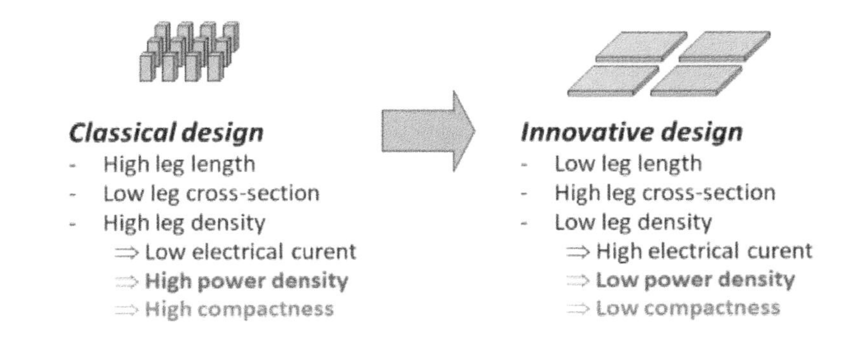

Figure 6.14. *Proposal for an innovative design of thermoelectric modules in order to improve their performance, in the case of low space constraints*

The reader will find an alternative design aimed at maximizing the cooling capacity for a reduced volume, in the case of a high power density application in Cheng and Lin (2005).

6.2.5. *Overall performance of thermoelectric systems*

As for any heat pump system, the thermal coupling between the TE system and its heat reservoirs (hot and cold) via heat exchangers also deserves special attention, as it involves a change in the junction temperatures, due to thermal access resistances between the system and its heat reservoirs (shown in Figure 6.11).

The thermal resistances, denoted respectively R_h^{th} at the hot side and R_c^{th} at the cold side, of the TE system reflect the resistance to heat transfer between the system and its heat reservoirs.

$$Q_h = \frac{T_h - {T_h}^S}{{R^{th}}_h} \quad [6.46]$$

$$Q_c = \frac{{T_c}^S - T_c}{{R^{th}}_c} \quad [6.47]$$

where T_h and T_c are the hot and cold junction temperatures respectively and $T_h{}^S$ and $T_c{}^S$ are the hot and cold heat reservoir temperatures respectively (as shown in Figure 6.11).

Since the temperature levels of the heat reservoirs are generally known and imposed by the intended application, the temperatures of the TE junctions are finally floating variables imposed by the operating conditions (electrical current and resulting thermal powers) and the reservoir temperatures, as described by the previous equations. Solving the system of coupled equations (equations [6.36]–[6.38] and [6.46]–[6.47]) is then necessary to determine the junction temperatures and the overall performance of the thermoelectric system.

The influence of this thermal coupling on the conversion efficiency of the system is discussed in Flores-Niño et al. (2015). It should be noted here that this influence is greater since the thermoelectric modules are generally sized for high power densities (high thermal powers distributed over small surfaces). The sizing of the heat exchangers, or heat sinks, associated with the module therefore proves to be critical due to its non-negligible influence on the overall performance of the system. An illustration of the performances (cooling power and conversion efficiency) according to the thermal resistances is presented in Figure 6.15.

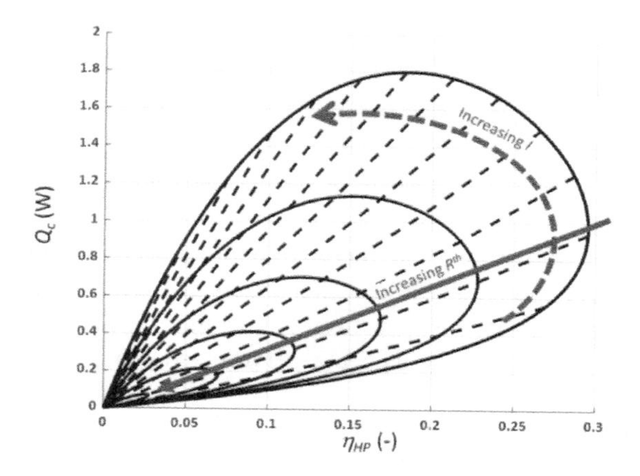

Figure 6.15. *Cooling thermal power Q_c and conversion efficiency η_{HP} of a Bi_2Te_3 thermoelectric module (S = 230 10^{-6} V.K^{-1}; λ = 1.6 W.m^{-1}.K^{-1}; ρ = 1.4 10^{-5} Ω.m; L = 5 10^{-3} m; A = 1 10^{-6} m^2), for electrical currents I varying between 0.5 and 1.7 A (in steps of 0.1 A) and thermal resistances $R_h^{th} = R_c^{th}$ between 0 and 1,000 K.W^{-1} (in steps of 200 K.W^{-1}), for reservoir temperatures $T_h^S = 350\ K$ and $T_c^S = 300\ K$. Solid lines are drawn for constant thermal resistances, and dotted lines for constant electrical currents*

The decrease in performance (heating power and conversion efficiency) induced by the presence of thermal resistances between the system and its heat reservoirs can then render the system inoperative ($Q_f < 0$ and $\eta_{HP} < 0$). Particular attention must therefore be paid to the design of the heat exchangers associated with the thermoelectric module. It should be noted that the intensification of heat exchanges generally involves an increase in pressure drops in the exchanger, resulting in overconsumption of the components involved for the circulation (pump/fan), thus contributing to reduce the overall performance of the system (Ramousse 2016).

The influence of additional thermal contact resistances between the thermoelectric module and its associated heat exchangers, resulting from imperfect thermal contact between the elements, is discussed in Xuan (2003).

An additional alternative can be considered, if the system compactness is not a strong constraint: increasing the spacing between the thermoelectric legs makes it possible to reduce the thermal power densities and thus ease heat exchange in the heat exchangers. However, this proposal encounters a technology challenge related to the implementation of the thermoelectric material, due to the strong thermal and mechanical constraints, caused by the differential thermal expansions of the different materials used in a module ("*n*" and "*p*" legs, electrical junctions, ceramics, heat exchanger, etc.).

6.2.6. *Thermodynamic analysis of irreversibilities*

The performance losses in conversion systems, compared to the ideal performance of reversible Carnot systems, are due to the different irreversibilities in the system. These irreversibilities are characterized by the entropy generation, evaluated using a second-law analysis. The actual conversion efficiency is then expressed as a function of the total entropy generation S_{gen}^{tot}, as follows:

$$\eta_{HP} = \eta^{C}{}_{HP} - \frac{S_{gen}^{tot} T_c}{Q_h} \qquad [6.48]$$

with S_{gen}^{tot} being positive.

The different contributions to the total entropy generation can be dissociated according to their origin. It is thus possible to distinguish the internally generated entropy (endo-irreversibilities) from those of external origin (exo-irreversibilities), linked to the heat exchangers associated with the thermoelectric module.

6.2.6.1. *Endo-irreversibilities*

The internal irreversibilities of the thermoelectric modules are due to the transfer phenomena involved in these systems, introduced previously. In this section, an ideal coupling to the heat reservoirs is assumed (exo-reversible system), implying null thermal resistances between the system and its heat reservoirs.

A detailed thermodynamic analysis of thermoelectric phenomena from the material scale to the system scale can be found in Goupil et al. (2011) and Apertet et al. (2013). Onsager reciprocal relations, discussed in Callen (1948), make it possible to affirm that the Seebeck effect, related to the coupling between electronic flows and heat flows, is a reversible phenomenon, which therefore does not contribute to internal irreversibilities in thermoelectric materials. The entropy generation, and the associated decreases in performance, are thus only linked to the resistive phenomena of charge and heat transfers in the material (thermal and electrical conductions). These two contributions to the internal entropy generation in thermoelectric systems can thus be written as follows:

$$S_{gen}^{Cond} = NK\Delta T\left(\frac{1}{T_c} - \frac{1}{T_h}\right) \quad [6.49]$$

$$S_{gen}^{Joule} = NRI^2\left(\frac{1}{T_c} - \frac{1}{T_h}\right) \quad [6.50]$$

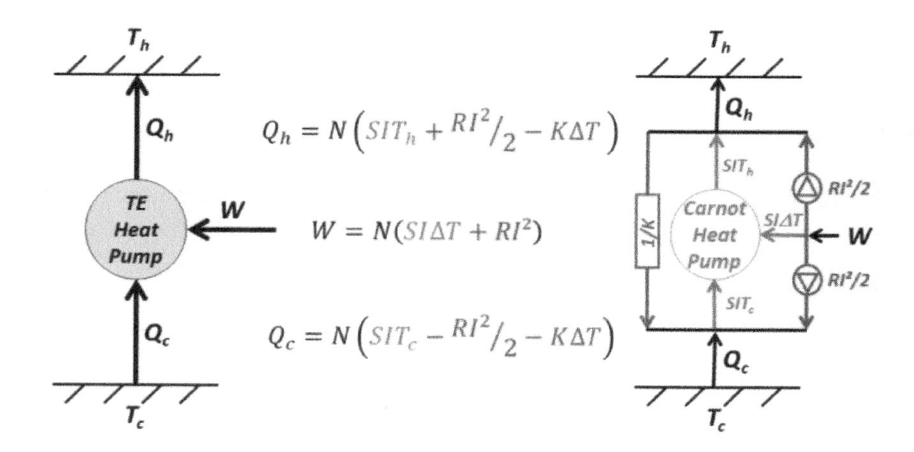

Figure 6.16. *Equivalent representation of a thermoelectric system by electrical analogy (intensity: energy flux; potential: temperature)*

Following this observation, a TE system can then be represented as a reversible (Carnot) system coupled to the irreversible phenomena of thermal conduction and Joule effect related to electrical conduction, as proposed in Figure 6.16 (Ramousse 2016).

6.2.6.2. *Exo-irreversibilites*

In addition to the internal irreversibilities of the TE system, it is necessary to take into account the external irreversibilities (exo-irreversibilities) associated with the thermal coupling of the heat pump systems to its heat reservoirs. The presence of thermal resistances between the system and its heat reservoirs, as presented in Figure 6.11, implies additional irreversibilities, thus resulting in performance losses as illustrated in Figure 6.15. The corresponding entropy generation results from the heat transfer between the thermoelectric system and the associated heat reservoirs, via the thermal resistances R_h^{th} and R_c^{th} of the heat exchangers, at the hot and cold sides respectively. It is thus expressed as a function of the thermal access resistances and of the temperatures of the heat reservoirs and of the thermoelectric junctions, as follows (Ramousse 2016):

$$S_{gen}^{Rth} = \frac{({T_h}^S - T_h)^2}{{R^{th}}_h T_h {T_h}^S} + \frac{({T_c}^S - T_c)^2}{{R^{th}}_c T_c {T_c}^S} \qquad [6.51]$$

The reduction of the thermal resistances thus makes it possible to reduce the associated irreversibilities (exo-irreversibilities). The thermal efficiency of heat exchangers therefore deserves to be taken into account for the design of an efficient thermoelectric device. It should be noted here that these contributions to performance degradation of thermoelectric systems can quickly become significant (especially for high power densities) and therefore require special attention in the design stage of thermoelectric systems, by integrating the design of heat exchangers associated with the system in a systemic approach. These considerations are in line with the proposal for the implementation of low power density thermoelectric modules mentioned above, with the aim of reducing its impact on the overall performance of the system. An illustration of the coupled influence between the overall design of the thermoelectric system and its associated heat exchangers is presented in Ramousse (2016).

In addition, viscous dissipation phenomena in the heat exchangers, resulting in additional electrical consumption of the circulation components, imply an additional contribution to the total entropy generation of the system. Although this generally remains negligible, it cannot be ruled out because it contributes to defining a compromise between thermal efficiency and electrical consumption, depending on the fluid flow used in the heat exchanger. This contribution can significantly penalize the overall performance of the system especially at very high Reynolds number (turbulent regime).

6.2.7. *Integration and management*

Finally, different management strategies for TE systems can be considered, to adjust to operating conditions (heat reservoir temperatures and/or thermal demand) that could be variable or transient. These control strategies can simply aim to adapt the electrical current supplied according to the operating conditions encountered, in order to maintain maximum performance at all times (in the case of temporal variations). This then requires the implementation of a controllable electrical charge. In the case of TE systems made up of several TE modules, it may also be appropriate to consider fluidic management of heat transfer fluids, in addition to the electrical management of the system (David et al. 2012). By modifying the thermal association (cascade or parallel association) between the different thermoelectric modules and the number of active modules, it is possible to maintain system performance independently of the power demand, as presented in Figure 6.17.

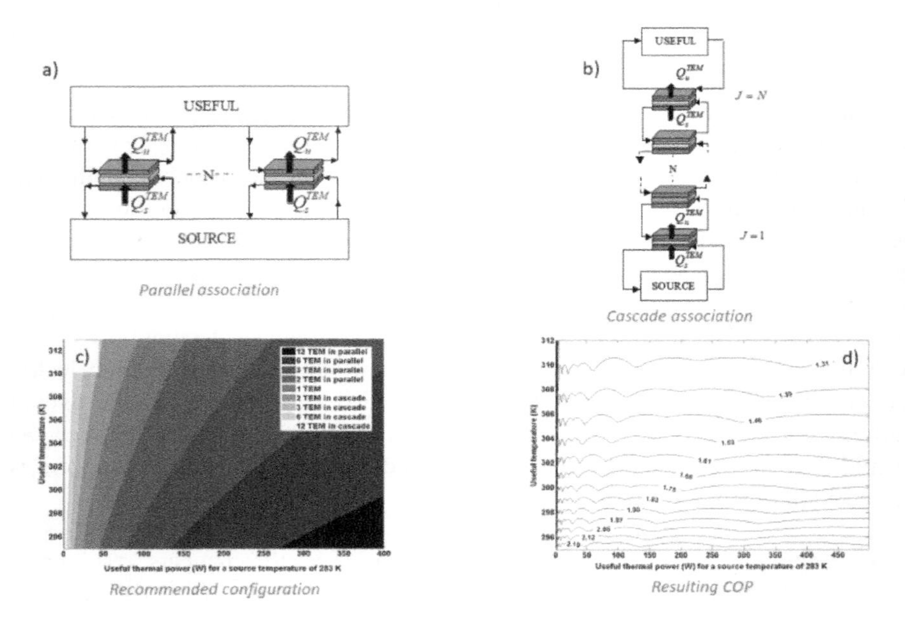

Figure 6.17. *Thermal association of thermoelectric modules a) in parallel, b) in cascade. c) Recommended association configuration and d) resulting COP as a function of the useful effect temperature T_h and thermal power Q_h, for a low temperature of T_c = 283 K*

Even if the increase in performance of such strategies is not always significant, it has been shown that an in-depth study of the operating conditions, depending on the implementation, could make it possible to significantly reduce the number of TE modules necessary for a TE residential heat pump-type application (Kim et al. 2014). Improving the performance of TE systems by the application of a pulsed current has been demonstrated in Snyder et al. (2002). The reader may refer to the following references for additional information on the use of this approach (Yang et al. 2005; Shen et al. 2012; Gao et al. 2017).

6.3. Applications

The implementation modularity of TE devices makes it possible to envision a wide field of applications. In addition, these devices operate without light and without removable parts, limiting noise generation and offering great stability, thus allowing their integration in very constrained environments or for very specific applications where other energy conversion devices cannot be considered. However, the still limited performance of these systems currently limits them to niche applications, such as in the aerospace, military, scientific and medical fields. In the civil field, the main areas of application of thermoelectric systems can be classified according to the following categories: cooling of electronic and optical components, domestic refrigerators, building applications, automotive cooling and solar cooling. The devices currently marketed or under study around ambient temperatures are mostly made from solid Bi_2Te_3, due to implementation process issues. The reader may refer to the following summary articles for additional information on the applications, their perspectives both from the material and system points of view and the associated modeling tools (Enescu and Virjoghe 2014; Zhao and Tan 2014; Pourkiaei et al. 2019). In addition, the review article Champier (2017) focuses on the generator mode.

6.3.1. *Cooling of electronic and optical components*

Maintaining the temperature of electronic and optical components, which are often responsible for the release of significant heat, is essential to preserve these sensitive components over the long term. For the majority of electronic components (including in particular computer processors and integrated circuits), a maximum temperature of 85°C must be respected. Due to the high power densities released (constantly increasing), conventional cooling techniques by convection (air or water) are often no longer sufficient and therefore require the implementation of active cooling systems. In this context, the cooling of electronic and optical components can thus be carried out using TE devices which can easily be integrated in these very constrained environments. The compactness and high reliability of TE systems are important advantages in this case. For these reasons, devices for cooling high-power microelectronic components by TE effects are currently widely available on the market.

In addition, TE systems are also used to ensure the temperature control of optical devices requiring a constant and reliable operating temperature, using appropriate regulation. The TE device is then alternately used to heat or cool the optical component, depending on the desired reference temperature. The performance challenges of these systems are mainly related to heat exchanges with the environment (generally by forced convection), as the TE devices are generally associated directly with the electronic component by direct thermal contact. Due to the better exchange coefficients with water than with air, cooling by forced convection with water should be preferred to efficiently regulate high power densities when possible.

6.3.2. *Domestic refrigerator*

TE devices can be considered for domestic refrigeration applications. The objective here is to cool (or maintain at a low temperature) a low-volume enclosure, such as refrigerators (portable or not), portable coolers, can coolers and insulated picnic baskets. In the case of portable applications, a power supply by batteries or via photovoltaic solar panels will be necessary.

The performances recorded for TE devices (cooling COP between 0.3 and 0.5 for an indoor/outdoor temperature difference of 20 to 25°C) are not yet up to those obtained in the case of conventional vapor compression thermodynamic systems (about 2.5). For these reasons, their implementation is often preferred for low-power applications (less than 25 W). Although this performance remains limited, TE devices have the following significant advantages: absence of refrigerant fluid (with high global warming potential), silent, robust, compact and precise temperature control. In particular, it has been shown that the enclosure temperature during the on/off cycles of the compressor for a vapor compression system can vary by several degrees, which can be detrimental to the preservation of food. The prospects for improving performance are mainly related to heat exchanges with heat reservoirs (mostly achieved by forced convection with air), resulting in significant thermal resistances.

6.3.3. *Building applications: air conditioning, room cooling*

TE systems can be an alternative to thermodynamic vapor compression cycle cooling systems. Although the performances currently achieved by thermoelectric devices (cooling COP from 0.4 to 1.5) are still far behind conventional thermodynamic systems (cooling COP from 2.6 to 3), TE processes have certain additional advantages: absence of refrigerant fluid, absence of moving parts (therefore reduction of noise pollution), reduced maintenance and, above all, possible inversion of heat flows by simple electrical control to alternately cover heating needs in winter and cooling in summer. The performances reported, obviously depending on the temperature conditions studied, are also strongly dependent on the mode of integration of the TE devices. Studies have proposed the insertion of thermoelements in so-called active windows (Arenas Alonso, n.d.) to overcome the difficulties related to conventional vapor compression systems implementation. A cold power of 150 W has been reported, with a decrease in transparency of less than 20%. The integration of TE elements into walls has also been patented (Ait Ameur et al. 2014), in order to provide heating/cooling by radiative effect to the indoor atmosphere. In the latter case, the low thermal power densities involved make it possible to mitigate decreases in performance linked to the heat exchanges with the thermal reservoirs.

6.3.4. *Automotive cooling*

TE applications are also envisioned in the automotive field. Due to the absence of refrigerant fluid, their compactness, their reliability and their ability to meet heating and cooling needs, TE devices seem particularly suitable for automotive applications. In addition to indoor cooling applications (which can be located near passengers), TE devices could also be integrated into the seats in order to meet passenger comfort (heating/cooling). The American company Gentherm, formerly Amerigon, already markets such a device, the "Climate Control Seat" system (CCS), for the comfort of drivers of many car manufacturers (Gentherm n.d.). This market for TE systems, dedicated to passenger comfort using localized indoor temperature control, seems to be expanding, in line with the goals to reduce CO_2 emissions in the automobile industry.

6.3.5. *Autonomous solar cooling*

Finally, by combining the Seebeck (TE generator) and Peltier (TE heat pump) effects, different applications are encountered to provide cooling from solar thermal energy (Dai et al. 2003). In these applications, solar thermal energy is used as a heat source for the generator module, which then supplies the electrical power to the module operating in heat pump mode. The thermal energy to be released by the system is thus the sum of the heat released by the two units (generator and heat pump). A reduction of 10°C in a volume of 250 mL of water in 2 hours at an outside temperature of 31°C has been demonstrated on a prototype. The perspectives for improving these complex systems relate to the design of the two modules and the associated heat exchangers, using a systemic approach due to the strong coupling between all the components of the system. This type of low-power application makes it possible to meet cooling needs for a remote site, without requiring a battery.

6.4. References

Ait Ameur, M., Schoeffler, F., Fillatre, D., Goupil, C., Gascoin, F., Tedenac, J.C., Fraisse, G., Cornu, D., Ruiz-Theron, E., Ramousse, J. (2014). Thermoelectric core, thermoelectric structure including said core, method for manufacturing same and uses thereof. Patent No. EP2592668.

Alam, H. and Ramakrishna, S. (2017). Nanostructured thermoelectric materials: Current research and future challenges. *Nanotechnology for Energy Sustainability*, 507–546. https://doi.org/10.1002/9783527696109.ch22.

Apertet, Y. and Goupil, C. (2016). On the fundamental aspect of the first Kelvin's relation in thermoelectricity. *International Journal of Thermal Sciences*, 104, 225–227. https://doi.org/10.1016/j.ijthermalsci.2016.01.009.

Apertet, Y., Ouerdane, H., Goupil, C., Lecoeur, P. (2013). From local force-flux relationships to internal dissipations and their impact on heat engine performance: The illustrative case of a thermoelectric generator. *Physical Review E*, 88(2), 022137. https://doi.org/10.1103/PhysRevE.88.022137.

Arenas Alonso, A. (n.d.). Transparent active thermoelectric wall. Patent No. 2151381.

Balout, H., Boulet, P., Record, M.-C. (2017). Strain-induced electronic band convergence: Effect on the Seebeck coefficient of Mg2Si for thermoelectric applications. *Journal of Molecular Modeling*, 23(4), 130. https://doi.org/10.1007/s00894-017-3304-1.

Behnia, K. (2015). *Fundamentals of Thermoelectricity*. Oxford University Press [Online]. Available at: https://doi.org/10.1093/acprof:oso/9780199697663.001.0001.

Bell, L.E. (2008). Cooling, heating, generating power, and recovering waste heat with thermoelectric systems. *Science*, 321(5895), 1457–1461. https://doi.org/10.1126/science.1158899.

Benyahia, M., Ohorodniichuk, V., Leroy, E., Dauscher, A., Lenoir, B., Alleno, E. (2018). High thermoelectric figure of merit in mesostructured In0.25Co4Sb12 n-type skutterudite. *Journal of Alloys and Compounds*, 735, 1096–1104. https://doi.org/10.1016/j.jallcom.2017.11.195.

Bouyrie, Y., Candolfi, C., Pailhès, S., Koza, M.M., Malaman, B., Dauscher, A., Tobola, J., Boisron, O., Saviot, L., Lenoir, B. (2015). From crystal to glass-like thermal conductivity in crystalline minerals. *Physical Chemistry Chemical Physics*, 17(30), 19751–19758. https://doi.org/10.1039/C5CP02900G.

Callen, H.B. (1948). The application of onsager's reciprocal relations to thermoelectric, thermomagnetic, and galvanomagnetic effects. *Physical Review*, 73(11), 1349–1358. https://doi.org/10.1103/PhysRev.73.1349.

Champier, D. (2017). Thermoelectric generators: A review of applications. *Energy Conversion and Management*, 140, 167–181. https://doi.org/10.1016/j.enconman.2017.02.070.

Chen, G., Dresselhaus, M.S., Dresselhaus, G., Fleurial, J.-P., Caillat, T. (2003). Recent developments in thermoelectric materials. *International Materials Reviews*, 48(1), 45–66. https://doi.org/10.1179/095066003225010182.

Cheng, Y.-H. and Lin, W.-K. (2005). Geometric optimization of thermoelectric coolers in a confined volume using genetic algorithms. *Applied Thermal Engineering*, 25(17–18), 2983–2997. https://doi.org/10.1016/j.applthermaleng.2005.03.007.

Cowen, L.M., Atoyo, J., Carnie, M.J., Baran, D., Schroeder, B.C. (2017). Review–organic materials for thermoelectric energy generation. *ECS Journal of Solid State Science and Technology*, 6(3), N3080–N3088. https://doi.org/10.1149/2.0121703jss.

Dai, Y., Wang, R., Ni, L. (2003). Experimental investigation and analysis on a thermoelectric refrigerator driven by solar cells. *Solar Energy Materials and Solar Cells*, 77(4), 377–391. https://doi.org/10.1016/S0927-0248(02)00357-4.

Datta, S. (2005). *Quantum Transport: Atom to Transistor*. Cambridge University Press. https://doi.org/10.1017/CBO9781139164313.

Datta, S. (2018). *Lessons from Nanoelectronics*, Volume 5. World Scientific. https://doi.org/10.1142/10440-vol2.

David, B., Ramousse, J., Luo, L. (2012). Optimization of thermoelectric heat pumps by operating condition management and heat exchanger design. *Energy Conversion and Management*, 60, 125–133. https://doi.org/10.1016/j.enconman.2012.02.007.

Dávila Pineda, D. and Rezania, A. (eds) (2017). *Thermoelectric Energy Conversion*. Wiley. https://doi.org/10.1002/9783527698110.

Di Castro, C. and Raimondi, R. (2015). *Statistical Mechanics and Applications in Condensed Matter*. Cambridge University Press. https://doi.org/10.1017/CBO9781139600286.

Di Ventra, M. (2010) *Electrical Transport in Nanoscale Systems.* Cambridge University Press. doi: 10.1017/CBO9780511755606.

Dresselhaus, M.S., Chen, G., Tang, M.Y., Yang, R.G., Lee, H., Wang, D.Z., Ren, Z.F., Fleurial, J.-P., Gogna, P. (2007). New directions for low-dimensional thermoelectric materials. *Advanced Materials*, 19(8), 1043–1053. https://doi.org/10.1002/adma.200600527.

Eibl, O., Nielsch, K., Peranio, N., Völklein, F. (eds) (2015). *Thermoelectric Bi 2 Te 3 Nanomaterials*. Wiley-VCH Verlag GmbH & Co. KGaA. https://doi.org/10.1002/9783527672608.

El Oualid, S., Kogut, I., Benyahia, M., Geczi, E., Kruck, U., Kosior, F., Masschelein, P., Candolfi, C., Dauscher, A., Koenig, J.D. et al. (2021). Thermoelectric generators: high power density thermoelectric generators with skutterudites. *Advanced Energy Materials*, 11(19). doi: 10.1002/aenm.202170073.

Enescu, D. and Virjoghe, E.O. (2014). A review on thermoelectric cooling parameters and performance. *Renewable and Sustainable Energy Reviews*, 38, 903–916. https://doi.org/10.1016/j.rser.2014.07.045.

Fitriani, Ovik, R., Long, B.D., Barma, M.C., Riaz, M., Sabri, M.F.M., Said, S.M., Saidur, R. (2016). A review on nanostructures of high-temperature thermoelectric materials for waste heat recovery. *Renewable and Sustainable Energy Reviews*, 64, 635–659. https://doi.org/10.1016/j.rser.2016.06.035.

Flores-Niño, C., Olivares-Robles, M., Loboda, I. (2015). General approach for composite thermoelectric systems with thermal coupling: The case of a dual thermoelectric cooler. *Entropy*, 17(6), 3787–3805. https://doi.org/10.3390/e17063787.

Fraisse, G., Ramousse, J., Sgorlon, D., Goupil, C. (2013). Comparison of different modeling approaches for thermoelectric elements. *Energy Conversion and Management*, 65, 351–356. https://doi.org/10.1016/j.enconman.2012.08.022.

Freer, R., Ekren, D., Ghosh, T., Biswas, K., Qiu, P., Wan, S., Chen, L., Han, S., Fu, C., Zhu, T. et al. (2022). Key properties of inorganic thermoelectric materials – tables. *J. Phys. Energy*, 4(2), doi: 10.1088/2515-7655/ac49dc.

Gao, Y.-W., Lv, H., Wang, X.-D., Yan, W.-M. (2017). Enhanced peltier cooling of two-stage thermoelectric cooler via pulse currents. *International Journal of Heat and Mass Transfer*, 114, 656–663. https://doi.org/10.1016/j.ijheatmasstransfer.2017.06.102.

Gayner, C. and Amouyal, Y. (2020). Energy filtering of charge carriers: Current trends, challenges, and prospects for thermoelectric materials. *Advanced Functional Materials*, 30(18), 1901789. https://doi.org/10.1002/adfm.201901789.

Gentherm (n.d.) Automotive solutions [Online]. Available at: https://gentherm.com/en/solutions/automotive/seat-comfort.

Goldsmid, H.J. (2016). *Introduction to Thermoelectricity*, vol. 121. Springer. https://doi.org/10.1007/978-3-662-49256-7.

Goupil, C. (ed.) (2016). *Continuum Theory and Modeling of Thermoelectric Elements*. Wiley. https://doi.org/10.1002/9783527338405.

Goupil, C., Seifert, W., Zabrocki, K., Müller, E., Snyder, G.J. (2011). Thermodynamics of thermoelectric phenomena and applications. *Entropy*, 13(8), 1481–1517. https://doi.org/10.3390/e13081481.

Guthmann, C., Diu, B., Lederer, D., Roulet, B. (2007) *Thermodynamique*. Éditions Hermann.

Hammond, C. (2015). *The Basics of Crystallography and Diffraction*. Oxford University Press. https://doi.org/10.1093/acprof:oso/9780198738671.001.0001.

Hao, S., Dravid, V.P., Kanatzidis, M.G., Wolverton, C. (2019). Computational strategies for design and discovery of nanostructured thermoelectrics. *npj Computational Materials*, 5(1), 58. https://doi.org/10.1038/s41524-019-0197-9.

He, J., Kanatzidis, M.G., Dravid, V.P. (2013). High performance bulk thermoelectrics via a panoscopic approach. *Materials Today*, 16(5), 166–176. https://doi.org/10.1016/j.mattod.2013.05.004.

He, R., Schierning, G., Nielsch, K. (2018). Thermoelectric devices: A review of devices, architectures, and contact optimization. *Advanced Materials Technologies*, 3(4), 1700256. https://doi.org/10.1002/admt.201700256.

Heremans, J.P., Wiendlocha, B., Chamoire, A.M. (2012). Resonant levels in bulk thermoelectric semiconductors. *Energy & Environmental Science*, 5(2), 5510–5530. https://doi.org/10.1039/C1EE02612G.

Hong, M., Lyv, W., Li, M., Xu, S., Sun, Q., Zou, J., Chen, Z.-G. (2020). Rashba effect maximizes thermoelectric performance of GeTe derivatives. *Joule*, 4(9), 2030–2043. https://doi.org/10.1016/j.joule.2020.07.021.

Hüfner, S. (2003). *Photoelectron Spectroscopy*. Springer. https://doi.org/10.1007/978-3-662-09280-4.

Ioffe, A.F., Stil'bans, L.S., Iordanishvili, E.K., Stavitskaya, T.S., Gelbtuch, A., Vineyard, G. (1959). Semiconductor thermoelements and thermoelectric cooling. *Physics Today*, 12(5), 42–42. https://doi.org/10.1063/1.3060810.

Jaziri, N., Boughamoura, A., Müller, J., Mezghani, B., Tounsi, F., Ismail, M. (2020). A comprehensive review of thermoelectric generators: Technologies and common applications. *Energy Reports*, 6, 264–287. https://doi.org/10.1016/j.egyr.2019.12.011.

Jiang, Q., Yang, J., Liu, Y., He, H. (2016). Microstructure tailoring in nanostructured thermoelectric materials. *Journal of Advanced Dielectrics*, 6(01), 1630002. https://doi.org/10.1142/S2010135X16300024.

Jood, P., Male, J.P., Anand, S., Matsushita, Y., Takagiwa, Y., Kanatzidis, M.G., Snyder, G.J., Ohta, M. (2020). Na doping in PbTe: Solubility, band convergence, phase boundary mapping, and thermoelectric properties. *Journal of the American Chemical Society*, 142(36), 15464–15475. https://doi.org/10.1021/jacs.0c07067.

Kaviany, M. (2008). *Heat Transfer Physics*. Cambridge University Press. https://doi.org/10.1017/CBO9780511754586.

Khire, R.A., Messac, A., Van Dessel, S. (2005). Design of thermoelectric heat pump unit for active building envelope systems. *International Journal of Heat and Mass Transfer*, 48(19–20), 4028–4040. https://doi.org/10.1016/j.ijheatmasstransfer.2005.04.028.

Kim, Y.W., Ramousse, J., Fraisse, G., Dalicieux, P., Baranek, P. (2014). Optimal sizing of a thermoelectric heat pump (THP) for heating energy-efficient buildings. *Energy and Buildings*, 70, 106–116. https://doi.org/10.1016/j.enbuild.2013.11.021.

Lee, H. (ed.) (2016). *Thermoelectrics: Design and Materials*. John Wiley & Sons, Ltd. https://doi.org/10.1002/9781118848944.

Lee, K.H. and Kim, O.J. (2007). Analysis on the cooling performance of the thermoelectric micro-cooler. *International Journal of Heat and Mass Transfer*, 50(9–10), 1982–1992. https://doi.org/10.1016/j.ijheatmasstransfer.2006.09.037.

Lee, K.H., Kim, S., Kim, H.-S., Kim, S.W. (2020). Band convergence in thermoelectric materials: Theoretical background and consideration on Bi–Sb–Te Alloys. *ACS Applied Energy Materials*, 3(3), 2214–2223. https://doi.org/10.1021/acsaem.9b02131.

Leong, C.F., Usov, P.M., D'Alessandro, D.M. (2016). Intrinsically conducting metal–organic frameworks. *MRS Bulletin*, 41(11), 858–864. https://doi.org/10.1557/mrs.2016.241.

Li, X., Sheng, Y., Wu, L., Hu, S., Yang, J., Singh, D.J., Yang, J., Zhang, W. (2020). Defect-mediated Rashba engineering for optimizing electrical transport in thermoelectric BiTeI. *npj Computational Materials*, 6(1), 107. https://doi.org/10.1038/s41524-020-00378-4.

Lin, Y., Wood, M., Imasato, K., Kuo, J.J., Lam, D., Mortazavi, A.N., Slade, T.J., Hodge, S.A., Xi, K., Kanatzidis, M.G. et al. (2020). Expression of interfacial seebeck coefficient through grain boundary engineering with multi-layer graphene nanoplatelets. *Energy & Environmental Science*, 13(11), 4114–4121. https://doi.org/10.1039/D0EE02490B.

Liu, W., Tan, X., Yin, K., Liu, H., Tang, X., Shi, J., Zhang, Q., Uher, C. (2012). Convergence of conduction bands as a means of enhancing thermoelectric performance of n-type Mg2Si(1-x)Sn(x) solid solutions. *Physical Review Letters*, 108(16), 166601. https://doi.org/10.1103/PhysRevLett.108.166601.

Lory, P.-F., Pailhès, S., Giordano, V.M., Euchner, H., Nguyen, H.D., Ramlau, R., Borrmann, H., Schmidt, M., Baitinger, M., Ikeda, M. et al. (2017). Direct measurement of individual phonon lifetimes in the clathrate compound Ba7.81Ge40.67Au5.33. *Nature Communications*, 8(1), 491. https://doi.org/10.1038/s41467-017-00584-7.

Ma, Z., Wei, J., Song, P., Zhang, M., Yang, L., Ma, J., Liu, W., Yang, F., Wang, X. (2021). Review of experimental approaches for improving zT of thermoelectric materials. *Materials Science in Semiconductor Processing*, 121, 105303. https://doi.org/10.1016/j.mssp.2020.105303.

Maaß, H., Bentmann, H., Seibel, C., Tusche, C., Eremeev, S.V., Peixoto, T.R.F., Tereshchenko, O.E., Kokh, K.A., Chulkov, E.V., Kirschner, J. et al. (2016). Spin-texture inversion in the giant Rashba semiconductor BiTeI. *Nature Communications*, 7(1), 11621. https://doi.org/10.1038/ncomms11621.

Mao, J., Liu, W., Ren, Z. (2016a). Carrier distribution in multi-band materials and its effect on thermoelectric properties. *Journal of Materiomics*, 2(2), 203–211. https://doi.org/10.1016/j.jmat.2016.03.001.

Mao, J., Liu, Z., Ren, Z. (2016b). Size effect in thermoelectric materials. *npj Quantum Materials*, 1(1), 16028. https://doi.org/10.1038/npjquantmats.2016.28.

Marder, M.P. (2010). *Condensed Matter Physics*. John Wiley & Sons, Inc. https://doi.org/10.1002/9780470949955.

Marino, E.C. (2017). *Quantum Field Theory Approach to Condensed Matter Physics*. Cambridge University Press. https://doi.org/10.1017/9781139696548.

Mehdizadeh Dehkordi, A., Zebarjadi, M., He, J., Tritt, T.M. (2015). Thermoelectric power factor: Enhancement mechanisms and strategies for higher performance thermoelectric materials. *Materials Science and Engineering: R: Reports*, 97, 1–22. https://doi.org/10.1016/j.mser.2015.08.001.

Minnich, A.J., Dresselhaus, M.S., Ren, Z.F., Chen, G. (2009). Bulk nanostructured thermoelectric materials: Current research and future prospects. *Energy & Environmental Science*, 2(5), 466. https://doi.org/10.1039/b822664b.

Misra, S., Wiendlocha, B., Tobola, J., Fesquet, F., Dauscher, A., Lenoir, B., Candolfi, C. (2020). Band structure engineering in Sn 1.03 Te through an In-induced resonant level. *Journal of Materials Chemistry C*, 8(3), 977–988. https://doi.org/10.1039/C9TC04407H.

Narducci, D. (2017). Energy filtering and thermoelectrics: Artifact or artifice? *Journal of Nanoscience and Nanotechnology*, 17(3), 1663–1667. https://doi.org/10.1166/jnn.2017.13726.

Neophytou, N. (2020). *Theory and Simulation Methods for Electronic and Phononic Transport in Thermoelectric Materials*. Springer International Publishing. https://doi.org/10.1007/978-3-030-38681-8.

Nolas, G.S., Sharp, J., Goldsmid, H.J. (2001). *Thermoelectrics* (vol. 45). Springer. https://doi.org/10.1007/978-3-662-04569-5.

Ouyang, Z. and Li, D. (2016). Modelling of segmented high-performance thermoelectric generators with effects of thermal radiation, electrical and thermal contact resistances. *Scientific Reports*, 6(1), 24123. https://doi.org/10.1038/srep24123.

Pailhès, S., Euchner, H., Giordano, V.M., Debord, R., Assy, A., Gomès, S., Bosak, A., Machon, D., Paschen, S., de Boissieu, M. (2014). Localization of propagative phonons in a perfectly crystalline solid. *Physical Review Letters*, 113(2), 025506. https://doi.org/10.1103/PhysRevLett.113.025506.

Pailhès, S., Giordano, V.M., Euchner, H., Lory, P.-F., de Boissieu, M., Euchner, H. (2017). X-rays and neutrons spectroscopy for the investigation of individual phonons properties in crystalline and amorphous solids. In *Nanostructured Semiconductors: Amorphization and Thermal Properties*, Termentzidis, K. (ed.). Jenny Stanford Publishing. https://doi.org/10.1201/9781315364452.

Pan, Y., Lin, B., Chen, J. (2007). Performance analysis and parametric optimal design of an irreversible multi-couple thermoelectric refrigerator under various operating conditions. *Applied Energy*, 84(9), 882–892. https://doi.org/10.1016/j.apenergy.2007.02.008.

Pandel, D., Banerjee, M.K., Singh, A.K. (2021). A review of the Mg2(Si,Sn) alloy system as emerging thermoelectric material: Experimental and modeling aspects. *Journal of Electronic Materials*, 50(1), 25–51. https://doi.org/10.1007/s11664-020-08591-z.

Pei, Y., Shi, X., LaLonde, A., Wang, H., Chen, L., Snyder, G.J. (2011). Convergence of electronic bands for high performance bulk thermoelectrics. *Nature*, 473(7345), 66–69. https://doi.org/10.1038/nature09996.

Polozine, A., Sirotinskaya, S., Schaeffer, L. (2014). History of development of thermoelectric materials for electric power generation and criteria of their quality. *Materials Research*, 17(5), 1260–1267. https://doi.org/10.1590/1516-1439.272214.

Pourkiaei, S.M., Ahmadi, M.H., Sadeghzadeh, M., Moosavi, S., Pourfayaz, F., Chen, L., Pour Yazdi, M.A., Kumar, R. (2019). Thermoelectric cooler and thermoelectric generator devices: A review of present and potential applications, modeling and materials. *Energy*, 186, 115849. https://doi.org/10.1016/j.energy.2019.07.179.

Ramousse, J. (2016). Entropy analysis of thermoelectric heat pumps including multi-channel heat exchangers: Design considerations. *International Journal of Thermodynamics*, 19(2), 82. https://doi.org/10.5541/ijot.5000153630.

Ramousse, J. and Goupil, C. (2018). Chart for thermoelectric systems operation based on a ternary diagram for bithermal systems. *Entropy*, 20(9), 666. https://doi.org/10.3390/e20090666.

Ramousse, J., Sgorlon, D., Fraisse, G., Perier-Muzet, M. (2015). Analytical optimal design of thermoelectric heat pumps. *Applied Thermal Engineering*, 82, 48–56. https://doi.org/10.1016/j.applthermaleng.2015.02.042.

Rowe, D.M. (ed.) (2018). *Thermoelectrics Handbook*. CRC Press. https://doi.org/10.1201/9781420038903.

Sharma, P.A. and Sugar, J.D. (2014). Obstacles to applications of nanostructured thermoelectric alloys. *Frontiers in Chemistry*, 2. https://doi.org/10.3389/fchem.2014.00111.

Shen, L.M., Xiao, F., Chen, H.X., Wang, S.W. (2012). Numerical and experimental analysis of transient supercooling effect of voltage pulse on thermoelectric element. *International Journal of Refrigeration*, 35(4), 1156–1165. https://doi.org/10.1016/j.ijrefrig.2012.02.004.

Shi, X., Chen, L., Uher, C. (2016). Recent advances in high-performance bulk thermoelectric materials. *International Materials Reviews*, 61(6), 379–415. https://doi.org/10.1080/09506608.2016.1183075.

Sholl, D.S. and Steckel, J.A. (2009). *Density Functional Theory*. John Wiley & Sons, Inc. https://doi.org/10.1002/9780470447710.

Skipidarov, S. and Nikitin, M. (eds) (2016). *Thermoelectrics for Power Generation – A Look at Trends in the Technology*. InTech. https://doi.org/10.5772/62753.

Snyder, G. and Toberer, E.S. (2008). Complex thermoelectric materials. *Nature Materials*, 7(2), 105–114. https://doi.org/10.1038/nmat2090.

Snyder, G., Fleurial, J-P., Caillat, T. (2002). Supercooling of Peltier cooler using a current pulse. *Journal of Applied Physics*, 92(1564). https://doi.org/10.1063/1.1489713.

Thébaud, S., Adessi, C., Pailhès, S., Bouzerar, G. (2017). Boosting the power factor with resonant states: A model study. *Physical Review B*, 96(7), 075201. https://doi.org/10.1103/PhysRevB.96.075201.

Toberer, E.S., May, A.F., Snyder, G.J. (2010). Zintl chemistry for designing high efficiency thermoelectric materials. *Chemistry of Materials*, 22(3), 624–634. https://doi.org/10.1021/cm901956r.

Toberer, E.S., Zevalkink, A., Snyder, G.J. (2011). Phonon engineering through crystal chemistry. *Journal of Materials Chemistry*, 21(40), 15843. https://doi.org/10.1039/c1jm11754h.

Tritt, T.M. (ed.) (2004). *Thermal Conductivity*. Springer. https://doi.org/10.1007/b136496.

Turner, S.R., Pailhès, S., Bourdarot, F., Ollivier, J., Raymond, S., Keller, T., Sidis, Y., Castellan, J.-P., Lory, P.-F., Euchner, H. et al. (2021). Impact of temperature and mode polarization on the acoustic phonon range in complex crystalline phases: A case study on intermetallic clathrates. *Physical Review Research*, 3(1), 013021. https://doi.org/10.1103/PhysRevResearch.3.013021.

Wang, M., Dong, R., Feng, X. (2021). Two-dimensional conjugated metal–organic frameworks (2D c-MOFs): Chemistry and function for MOFtronics. *Chemical Society Reviews*, 50(4), 2764–2793. https://doi.org/10.1039/D0CS01160F.

Wiendlocha, B. (2016). Resonant levels, vacancies, and doping in Bi2Te3, Bi2Te2Se, and Bi2Se3 Tetradymites. *Journal of Electronic Materials*, 45(7), 3515–3531. https://doi.org/10.1007/s11664-016-4502-9.

Wu, L., Li, X., Wang, S., Zhang, T., Yang, J., Zhang, W., Chen, L., Yang, J. (2017). Resonant level-induced high thermoelectric response in indium-doped GeTe. *NPG Asia Materials*, 9(1), e343–e343. https://doi.org/10.1038/am.2016.203.

Xuan, X. (2003). Investigation of thermal contact effect on thermoelectric coolers. *Energy Conversion and Management*, 44(3), 399–410. https://doi.org/10.1016/S0196-8904(02)00062-6.

Yamashita, O. (2008). Effect of temperature dependence of electrical resistivity on the cooling performance of a single thermoelectric element. *Applied Energy*, 85(10), 1002–1014. https://doi.org/10.1016/j.apenergy.2008.02.011.

Yang, R., Chen, G., Ravi Kumar, A., Snyder, G.J., Fleurial, J.-P. (2005). Transient cooling of thermoelectric coolers and its applications for microdevices. *Energy Conversion and Management*, 46(9–10), 1407–1421. https://doi.org/10.1016/j.enconman.2004.07.004.

Yang, J., Xi, L., Qiu, W., Wu, L., Shi, X., Chen, L., Yang, J., Zhang, W., Uher, C., Singh, D.J. (2016). On the tuning of electrical and thermal transport in thermoelectrics: An integrated theory–experiment perspective. *npj Computational Materials*, 2(1), 15015. https://doi.org/10.1038/npjcompumats.2015.15.

Yoshioka, D. (2007) *Statistical Physics: An Introduction*, Springer. https://doi.org/10.1007/978-3-540-28606-6.

Yu, P.Y. and Cardona, M. (2010). *Fundamentals of Semiconductors*. Springer. https://doi.org/10.1007/978-3-642-00710-1.

Yuan, H., Liu, Z., Xu, G., Zhou, B., Wu, S., Dumcenco, D., Yan, K., Zhang, Y., Mo, S.-K., Dudin, P. et al. (2016). Evolution of the valley position in bulk transition-metal chalcogenides and their monolayer limit. *Nano Letters*, 16(8), 4738–4745. https://doi.org/10.1021/acs.nanolett.5b05107.

Yusuf, A. and Ballikaya, S. (2020). Modelling a segmented skutterudite-based thermoelectric generator to achieve maximum conversion efficiency. *Applied Sciences*, 10(1), 408. https://doi.org/10.3390/app10010408.

Zhang, X. and Zhao, L.-D. (2015). Thermoelectric materials: Energy conversion between heat and electricity. *Journal of Materiomics*, 1(2), 92–105. https://doi.org/10.1016/j.jmat.2015.01.001.

Zhao, D. and Tan, G. (2014). A review of thermoelectric cooling: Materials, modeling and applications. *Applied Thermal Engineering*, 66(1–2), 15–24. https://doi.org/10.1016/j.applthermaleng.2014.01.074.

Zhao, L.-D., He, J., Berardan, D., Lin, Y., Li, J.-F., Nan, C.-W., Dragoe, N. (2014). BiCuSeO oxyselenides: New promising thermoelectric materials. *Energy & Environmental Science*, 7(9), 2900–2924. https://doi.org/10.1039/C4EE00997E.

Zhou, J., Liao, B., Qiu, B., Huberman, S., Esfarjani, K., Dresselhaus, M.S., Chen, G. (2015). Ab initio optimization of phonon drag effect for lower-temperature thermoelectric energy conversion. *Proceedings of the National Academy of Sciences*, 112(48), 14777–14782. https://doi.org/10.1073/pnas.1512328112.

Zhu, H., Mao, J., Li, Y., Sun, J., Wang, Y., Zhu, Q., Li, G., Song, Q., Zhou, J., Fu, Y. et al. (2019). Discovery of TaFeSb-based half-Heuslers with high thermoelectric performance. *Nature Communications*, 10(1), 270. https://doi.org/10.1038/s41467-018-08223-5.

List of Authors

Jocelyn BONJOUR
CETHIL, INSA Lyon
Villeurbanne
France

Michel FEIDT
Université de Lorraine
LEMTA, CNRS
Nancy
France

Florine GIRAUD
Lafset
CNAM Paris
France

Philippe HABERSCHILL
CETHIL, INSA Lyon
Villeurbanne
France

Stéphane PAILHÈS
ILM UMR 5306
UCBL, CNRS, Université de Lyon
Villeurbanne
France

Julien RAMOUSSE
LOCIE UMR 5271
CNRS, Université Savoie Mont Blanc
Le Bourget du Lac
France

Rémi REVELLIN
CETHIL, INSA Lyon
Villeurbanne
France

Romuald RULLIÈRE
CETHIL, INSA Lyon
Villeurbanne
France

Monica SIROUX
ICUBE UMR 7357
INSA Strasbourg
CNRS, Université de Strasbourg
France

Index

H, I

L, M

O, P

R, S, T

Printed and bound by CPI Group (UK) Ltd, Croydon, CR0 4YY

06/08/2023

03243866-0001